Real-Life Math
PROBABILITY

SECOND EDITION

WALCH PUBLISHING

1   2   3   4   5   6   7   8   9   10

ISBN 978-0-8251-6321-0

Copyright © 1998, 2007

J. Weston Walch, Publisher

P. O. Box 658 • Portland, Maine 04104-0658

www.walch.com

Printed in the United States of America

Table of Contents

How to Use This Series v

Foreword . . . . *vi*

Basics

1. Probability All Around Us 1

2. Counting Choices 4

3. Independent Events and the Multiplication Rule 6

4. Factorials 9

5. Ways to Seat Guests: Permutations 12

6. Ways to Choose Lunch: Combinations 14

7. Scrambled Word Puzzles 18

Real-World Examples

8. Trials: Single Coin Toss 21

9. Analysis: Single Coin Toss 24

10. Two Dice: When Will Doubles Occur? 27

11. Two Dice: Sums 30

12. Analysis: Doubles or Sums 33

13. Password Possibilities 36

14. Setting Up a Basketball Tournament 39

15. Winning a Best-of-7-Game Series 42

16. What Is a Fair Price? 46

17. Number of Boys and Girls in a Family 49

18. Chance of a Hitting Streak 52

19. Two People in a Group With the Same Birthday 55

20. Medical Testing 59

21. Lotto Games: Winning the Big One? 62

Table of Contents

22. Waiting for the Academy Award® 65

23. Test for ESP 68

24. Hold Time 71

25. Parking at the Mall 76

26. The 3-Door Problem 78

Appendix

Random Number Tables 82–83

Mock Lotto Tickets 84

Random Date Template 85

How to Use This Series

The *Real-Life Math* series is a collection of activities designed to put math into the context of real-world settings. This series contains math appropriate for pre-algebra students all the way up to pre-calculus students. Problems can be used as reminders of old skills in new contexts, as an opportunity to show how a particular skill is used, or as an enrichment activity for stronger students. Because this is a collection of reproducible activities, you may make as many copies of each activity as you wish.

Please be aware that this collection does not and cannot replace teacher supervision. Although formulas are often given on the student page, this does not replace teacher instruction on the subjects to be covered. Teaching notes include extension suggestions, some of which may involve the use of outside experts. If it is not possible to get these presenters to come to your classroom, it may be desirable to have individual students contact them.

We have found a significant number of real-world settings for this collection, but it is not a complete list. Let your imagination go, and use your own experience or the experience of your students to create similar opportunities for contextual study.

Foreword

You've seen it happen many times—a player in a dice game claims she is "due" for doubles; strangers discover that they have a mutual acquaintance and think that this must be more than a chance meeting; a friend plays the lottery obsessively or enters online contests with a persistent dream of winning. All these behaviors reflect how people perceive probability in daily life. People who lack an accurate sense of probability are easily drawn in by false claims and pseudoscience, are vulnerable to get-rich-quick schemes, and exhibit many of the behaviors mentioned above.

The modeling and measurement of probabilities are fundamentals of mathematics that can be applied to the world around us. Every event, every measurement, every game, every accident, and even the nature of matter itself is understood through probabilistic models, yet few people have a good grasp of the nature of probability. Even students who have taken typical mathematics courses are unlikely to acquire the mathematical skills necessary to apply probabilistic models to real-world situations.

This book will help fill the gaps. This collection of activities will supplement general math, pre-algebra, or algebra courses, and will add focus to a course dedicated to probability. The activities will help students develop the mathematical foundation needed to understand how probability works. There are two sets of activities: One set (Activities 1–7) focuses on certain mathematical basics that are needed to understand applied examples. The other set (Activities 8–26) provides real-world examples. The first set is important because students will be lost when they see a problem if they do not understand a few basic principles and do not have the ability to do basic calculations. The second set forms the bulk of the book, moving from simple experiments to more challenging applications.

After mastering the activities in this book, students will have tools to help them evaluate the probabilities of events they will encounter, and in the process, they will learn to make better decisions in life.

—Eric T. Olson

1. Probability All Around Us

Context

daily life

Math Topic

random events

Overview

Probabilistic events affect everyday life in innumerable ways. In this activity, students are asked to make observations about events that have a random component.

Objectives

Students will be able to:

- describe a random event

- state random and nonrandom properties of events

Materials

- one copy of the Activity 1 handout for each student

Teaching Notes

To get students thinking about probability, ask them to come up with a list of events, games, and other everyday occurrences that have a perceived random property.

Discuss the concept of random events. If students are asked to name events in which the results are random, they should be able to come up with coin flips, dice rolls,

playing card deals, lottery draws, and so on. These kinds of common events are certainly one type of phenomenon that can be considered, but many other phenomena have random components. In fact, nearly any observation of daily activities reveals some underlying randomness.

Emphasize that the outcome of a single random event or experiment of chance cannot be exactly predicted. The variability of a single event makes exact prediction of the outcome impossible. There is no way to know if a coin flip will yield heads or tails or if the weather will be rainy on a particular day six months from now. However, for a large set of repeated random events, there is a useful regularity. This regularity can be expressed by probability. Probabilities express the mathematical likelihood of each possible outcome of a random event. It must be strongly emphasized that probabilities may be measured only for large sets of repetitions. Using a corresponding probability model, mathematics may be used to make useful decisions about events having a random component.

Answers

1. The entries in the table on page 3 are meant to be concise. You might expound on the individual items in a class discussion. Answers will vary, but may include those given on the next page.

(continued)

1. Probability All Around Us

Event	Nonrandom Property	Random Property
weather	seasonal change, local climate	precipitation, temperature on specific days
car accidents	safe or unsafe driving practices	specific cars or conditions met on the road
class grades	amount of study and preparation	appearance of specific questions on tests
customers at mall	hours open, time of day	specific pattern of customer arrival
state lottery	decisions about games offered, prizes	numbers drawn or winning patterns on tickets

2. The term *random* describes events that have underlying variability. Single random events cannot be predicted absolutely. However, using probability and probability models, it is possible to predict the frequencies of outcomes in a large set of random events.

Extension Activity

Have students spend a day or more observing events with random properties and report back to class.

1. Probability All Around Us

After your third softball game in a row is rained out, you are talking with your teammates. One of them says, "Some things just seem to happen. We scheduled these games months ago, but who could have predicted so much rain?"

You ask, "I wonder if this is what my math teacher means by 'random events'?"

Then you all start thinking about how many situations in daily life seem to happen randomly.

1. The table below lists five common events that have both random and nonrandom aspects. Explain what is random and what is not random about each. Then come up with five examples of your own.

Event	Nonrandom Property	Random Property
1. weather		
2. car accidents		
3. class grades		
4. customers at mall		
5. state lottery		
6.		
7.		
8.		
9.		
10.		

2. Describe what is meant by *random event.* Explain the relationship between probability and randomness.

2. Counting Choices

Context

choices in daily life

Math Topic

multiplication

Overview

The most important mathematical concept for understanding probability is counting. More specifically, it is necessary to learn exactly what to count, along with the patterns involved in counting sometimes large numbers of possibilities. This activity is about counting through multiplication. The concept of probability is introduced for a set of equally likely outcomes.

Objectives

Students will be able to:

- count possible outcomes of random events

- assign probabilities of equally likely events

Materials

- one copy of the Activity 2 handout for each student

Teaching Notes

Students will discover how multiple choices of several different items can be multiplied to obtain the total number of choices. Explain to students that most of the work in

determining the mathematical probability of an event is in counting the possible outcomes. The whole set of possible outcomes is called the *sample space.*

Answers

1. a. $12 \times 8 \times 5 \times 2 = 960$

    b. Following the first sock you choose, there are 8 ways to choose an unmatched sock and 1 way to choose a matched sock on your second pick from the 9 remaining socks. So, you will choose a matched sock $1/9$ of the time and an unmatched sock $8/9$ of the time.

    c. To determine the number of outfits that won't match, multiply the two nonmatching sock pairs by half the shirt-and-pants combinations:
    $2 \times (12 \times 8) / 2 = 96$
    So, 96 out of 960 possible outfits will include socks that don't go with the outfit. On 1 out of 10 days, you will wear socks of the wrong color.

    d. Answers will vary.

2. a. There are $5 \times 3 \times 4$, or 60, possible meals.

    b. There are $3 \times 1 \times 3$, or 9, meals that you like. Your chance (probability) of getting a meal you like is only $9/60$, or 15%.

2. Counting Choices

Our days are filled with choices. Sometimes it's confusing to even think about them. Read the situations below and answer the questions that follow.

1. In the morning you are always in a hurry, so you randomly choose clothes to wear from your closet (shirt, pants, socks, and shoes). You hope that what you pick out is not mismatched.

    a. If you have 12 shirts, 8 pairs of pants, 5 pairs of socks (each pair a different color), and 2 pairs of shoes, how many different outfits are possible? (Luckily, your socks and shoes are always in pairs.)

    b. If your socks all became separated, how often would you randomly pick out matched socks? mismatched socks?

    c. If 3 out of 5 of your (matched) pairs of socks go with all of your shirt-and-pants combinations, and the other 2 pairs of socks go with half of your shirt-and-pants combinations, how often would you choose socks that don't match the rest of your outfit?

    d. How could you buy clothes so that you would have the largest number of outfits for the least amount of money?

2. When you go to a local restaurant, there are 5 choices of sandwiches, 3 choices of potatoes, and 4 choices of soft drinks on the menu.

    a. How many different meals are possible at this restaurant?

    b. You like only 3 of the 5 sandwiches and 1 of the 4 kinds of soft drinks, but all 3 kinds of potatoes. If your friend brings you a meal selected at random from the restaurant's menu, what is the chance (the probability) that you will like it?

3. Independent Events and the Multiplication Rule

Context

probability basics

Math Topic

independence

Overview

Most of the examples in this book involve events that are independent. This means that one outcome or result is not influenced by another. Two events are independent if the result of the first event has no effect on the result of the second. Two coin flips are clearly independent, but there are cases where independence is not so clear. If events are independent, then an important mathematical rule of probabilities applies— the multiplication rule.

Objectives

Students will be able to:

- identify events that are independent

- apply the multiplication rule to determine the probability of a sequence of independent events

Materials

- one copy of the Activity 3 handout for each student

- calculator

Teaching Notes

In the activity, students are asked to identify events that are independent and ones that are not independent. The criterion they should learn to look for is influence: Will one event influence the other? Ask students to propose other sets of events that are independent or not independent.

The remaining exercises are all applications of the multiplication rule for mathematical probabilities. Students should calculate the probability of the sequence of events occurring. A few of the items do not involve independent events. In these cases, students should write "not independent" instead of calculating the probability.

Answers

1. a. independent

    b. independent

    c. not independent (removal of the aces increases the probability of drawing kings)

    d. not independent (removing the first red ball increases the probability of drawing a white ball)

2. a. not independent

    b. $(1/6) \times (1/6) = 1/36$

    c. $(1/13) \times (1/13) = 1/169$

    d. $0.01 \times 0.01 = 0.0001$, or $1/10{,}000$

    (continued)

 Real-Life Math: Probability

3. Independent Events and the Multiplication Rule

Events are independent when one result is not influenced by another. Some events are clearly independent. Other cases are not so clear. The element to look for is influence. Will one event influence the other?

1. Decide whether or not the following sequences of events are independent. If they are, write **independent**. If they are not, write **not independent** and explain why they are not.

    a. flipping a coin and getting heads, then flipping it again and getting tails

    b. rolling doubles with two dice, then rolling doubles again

    c. drawing (and removing) 2 aces from a normal deck of cards, then drawing 2 kings

    d. drawing (and removing) 1 red ball from a bag initially containing 2 red balls and 3 white balls, then drawing 1 white ball

(continued)

3. Independent Events and the Multiplication Rule

2. Find the probability for each of the sets of events given below by using the multiplication rule if the events are independent. If they are not independent, write **not independent**.

   a. randomly choosing 1 person who is over 6 feet tall and weighs more than 250 pounds from a group of 100 people, among whom there are 13 people over 6 feet tall and 8 people who weigh more than 250 pounds

   b. rolling a 6 with one die, then rolling again and getting another 6

   c. drawing an ace from a normal deck of 52 cards, replacing the card, reshuffling, and drawing an ace again

   d. two separate machines, each 99% reliable (1% probability of breaking down), both breaking down

4. Factorials

Context

probability basics

Math Topic

discrete math

Overview

This activity focuses on a single mathematical operation: factorials. Many activities on probability use factorials, so it is critical that students recognize them and know how to work with them. Use this activity to be certain students have this knowledge before they move on to Activities 5 and 6.

Objectives

Students will be able to:

- recognize and compute factorials for nonnegative integers

- realize that factorials become large quickly

- evaluate simple mathematical expressions containing factorials

Materials

- one copy of the Activity 4 handout for each student

- calculator

Teaching Notes

Students may be baffled when they see the symbol "!" in a mathematical expression. Its meaning is simple: For an integer n, $n!$ means multiply all of the integers 1, 2, 3, . . . , n together to produce a result called n factorial. It's easy to calculate factorials up to about 11 with a simple calculator. Students should do this.

Have students evaluate expressions that contain factorials. The point here is to demonstrate that some expressions may be impossible to evaluate unless computations are performed in the correct order. The answers are actually simple to obtain. For example, $1001!/1000!$ would be unwieldy if you attempted to compute 1000! and 1001! individually. However, if students realize that $1001! = 1000! \times 1001$, then $1001!/1000! = (1000! \times 1001)/1000! = 1001$. It is essential to understand this kind of operation, especially when working with combinations in the following activities.

Answers

1. a. 1

    b. 1

    c. 2

    d. 6

    e. 720

    f. 362,880

(continued)

4. Factorials

g. 3,628,800

h. 39,916,800

2. a. 6

   b. 120

   c. 100

   d. 4950

   e. 30,240

   f. 35

Extension Activity

Have students use a simple 8-digit calculator and a scientific calculator to determine the largest factorial that will fit in the display window of each calculator. (8-digit calculator: 11! = 39,916,800; scientific calculator: 69! = 1.7112×10^{98})

4. Factorials

For an integer *n*, *n*! means multiply all of the integers 1, 2, 3, ... , *n* together to produce a result called *n* factorial (*Note:* 0! is defined to equal 1).

Example: $5! = 5 \times 4 \times 3 \times 2 \times 1$

1. Calculate each factorial below.

    a.   0! =

    b.   1! =

    c.   2! =

    d.   3! =

    e.   6! =

    f.   9! =

    g.   10! =

    h.   11! =

2. Evaluate each factorial expression below.

    a. $6!/5! =$

    b. $16!/(2! \times 14!) =$

    c. $100!/99! =$

    d. $100!/(2! \times 98!) =$

    e. $7! \times 3! =$

    f. $7!/(4! \times 3!) =$

5. Ways to Seat Guests: Permutations

Context

entertainment

Math Topic

discrete math

Overview

A permutation is an arrangement of a set of numbers, people, or objects in a row or other sequence. Once students understand how to take factorials, permutations are easy.

Objectives

Students will be able to:

- calculate the number of permutations of n distinct items

Materials

- one copy of the Activity 5 handout for each student

- calculator

Teaching Notes

The example of possible seating arrangements of guests around a table is used to help students understand the pattern involved when counting the number of possible arrangements of n distinct items into n identifiable positions. For the first position, any of the guests may be seated anywhere; hence there are n possibilities for this position. After the first guest has been seated, $(n - 1)$ possibilities are left for the second position, $(n - 2)$ for the third, $(n - 3)$ for the fourth, and so on. To get the total number of possibilities, multiply

$$n \times (n - 1) \times (n - 2) \times (n - 3) \times \ldots \times 1 = n!$$

So the total number of arrangements of n distinct items is $n!$.

If there is a group of k (where k is greater than or equal to 2) identical or indistinguishable items in the set of n, then there are $k!$ equivalent arrangements for each permutation. Therefore, the total number of arrangements will be $n!/k!$.

Answers

1.   a.   5

      b.   4

      c.   3

      d.   2

      e.   1

2. To find the total number of ways to fill all the seats, multiply the number of ways to fill each seat:

$$5 \times 4 \times 3 \times 2 \times 1$$

This is the expression 5! or 5 factorial.

3. $5! = 120, 6! = 720$

5. Ways to Seat Guests: Permutations

Imagine that you have 5 dinner guests. While you are thinking about seating arrangements, you realize that there are many different ways you could seat your guests. But is there a way to figure out exactly how many arrangements are possible? Follow the instructions for question 1 exactly and in order. People are "available" until they are crossed out.

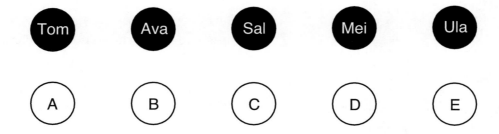

1. Choose a person to sit in seats A, B, C, D, and E one at a time and in order. Answer questions a–e one at a time as you fill in the name you have chosen for each seat in the appropriate circle.

   a. How many people are available to sit in seat A?

   b. How many people are available to sit in seat B?

   c. How many people are available to sit in seat C?

   d. How many people are available to sit in seat D?

   e. How many people are available to sit in seat E?

2. In question 1 above, you counted the number of persons available to fill each seat. How could you figure out the total number of ways there are to fill all the seats? Write this out as a mathematical expression. What is this mathematical expression called?

3. How many ways is it possible to arrange 5 people in 5 seats? 6 people in 6 seats?

6. Ways to Choose Lunch: Combinations

Context

entertainment

Math Topic

discrete math

Overview

Many problems in probability involve choosing combinations of numbers or items from a much larger group. Several activities in this book require an understanding of combinations; students should have a firm grasp of combinations before they attempt these activities.

Objectives

Students will be able to:

- calculate the number of combinations of n things chosen m at a time

Materials

- one copy of the Activity 6 handout for each student

- calculator

Teaching Notes

Emphasize that factorials and permutations are essential for understanding combinations. The math used is similar. Students should have mastered the computations in Activity 4, Factorials, and

Activity 5, Permutations, before going on to this activity.

Questions 1 and 2 of the activity allow students to develop a sense of combinations. They choose 2 food items from a group of 6 and write down the number of choices they had. Multiplying the number of choices for each pick gives the total number of ways to choose.

The concept of order is introduced in question 3. The total number of possible selections must be divided by the number of ways to arrange each selection to obtain the number of unique selections.

Finally, students are given a fast, easy-to-remember method for finding any combination's result. You may wish to introduce the formal mathematical procedure:

$$\binom{n}{m} = \frac{n!}{m!(n-m)!}$$

Answers

1. a. 6

    b. 5

2. Multiply the number of possibilities for each choice: $6 \times 5 = 30$.

3. Since there are 2 equivalent arrangements of each choice, there are $^{30}\!/_2$, or 15, unique choices. This represents enough combinations for a different lunch every weekday for

(continued)

6. Ways to Choose Lunch: Combinations

3 weeks. Be wary of students just multiplying 6×2 and not getting enough lunches.

4. a. $4 \times {}^3/_{2!} = 6$

   b. $3 \times {}^2/_{2!} = 3$

   c. $20 \times 19 \times 18 \times 17 \times 16 \times {}^{15}/_{6!} = 38{,}760$

   d. $6 \times 5 \times 4 \times 3 \times {}^2/_{5!} = 6$

Extension Activity

Have students bring in menus from a variety of restaurants. Ask them to rewrite this activity to reference the details of one of the menus (specific dishes, toppings, sides, etc.). Does the number of possible combinations have any relevance to managing a restaurant? Have students identify potential implications for planning, pricing, advertising, and so forth.

Real-Life Math: Probability

6. Ways to Choose Lunch: Combinations

You go to your favorite Mexican restaurant for the lunch special. You can choose any 2 different items for only $2.99. There are 6 items to choose from. You want to have a different combination every day, Monday through Friday, for the next 3 weeks. Figure out how this would be possible.

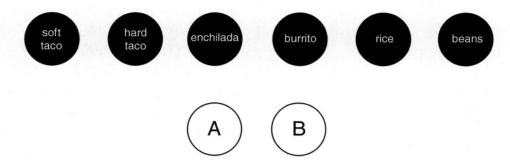

1. a. How many items are available to choose from for item A? Choose any one of the items, cross it out, and write its name across circle A.

    b. How many items are left to choose from for item B? Choose any one of the available items, cross it out, and write its name across circle B.

2. How would you calculate the total number of ways to fill circles A and B? How many ways can you fill the circles?

3. You are not concerned about the order of A and B. In other words, both taco/burrito and burrito/taco are considered as only one possible lunch. How many ways are there to choose lunch? Is this enough for 3 weeks of lunches?

(continued)

6. Ways to Choose Lunch: Combinations

4. There is an easy way to calculate the number of ways a certain number of items may be chosen from a larger group. First, determine the number of items you will choose. Then multiply the highest numbers together, until you have a number of factors equal to the number of items. For example, if you choose 3 objects from a group of 12, you would multiply $12 \times 11 \times 10$, the three highest numbers in the group. Then, if you are not concerned about the order of the objects chosen, divide by the number of ways you can arrange them. This is the number of permutations, equal to the factorial of the number chosen. In this case, the answer would be $(12 \times 11 \times 10) / 3!$, or 220 ways to select 3 objects from a group of 12. If you are not concerned about order, how many ways are there to:

a. select 2 from a group of 4?

b. select 2 from a group of 3?

c. select 6 from a group of 20?

d. select 5 from a group of 6?

7. Scrambled Word Puzzles

Context

puzzles and games

Math Topic

permutations and combinations

Overview

A common kind of word puzzle involves unscrambling several words with five or six letters and then using letters found in certain positions of the resulting words to decode a clever phrase. This activity gives students a chance to use their knowledge of permutations to solve puzzles of this kind.

Objectives

Students will be able to:

- recognize and apply knowledge of permutations to the solution of a scrambled word puzzle

Materials

- one copy of the Activity 7 handout for each pair of students

Teaching Notes

This activity is meant to be fun; it should be used as a break from the normal routine. But there are also learning objectives that your students should have achieved and should be able to apply in this activity. Specifically, they should be able to answer

the questions concerning the total number of possible permutations of any number of distinct letters. Some creative thought should allow them to figure out the permutations possible if letters are restricted to certain positions or if letters repeat.

Answers

1. There are 4 distinct letters, so there are 4!, or 24, permutations: *APST, APTS, ATPS, ATSP, ASPT, ASTP, PAST, PATS, PSTA, PSAT, PTSA, PTAS, SAPT, SATP, SPAT, SPTA, STAP, STPA, TAPS, TASP, TPAS, TPSA, TSAP, TSPA*. There are four actual words: *past, pats, spat,* and *taps*.

2. a. 5!, or 120

    b. 6!, or 720

    c. There are 3 identical letters, so for each permutation there are 3!, or 6, equivalents. The total is thus $^{120}/_6$ or 20 possible arrangements

    d. There are 3!, or 6, ways to arrange the letters in each group, so there are $6 \times 6 = 36$ ways to arrange these letters, significantly fewer than the 720 possible ways when all 6 letters are separate rather than in groups.

3. RANDOM, DEDUCE, SAMPLES, PREDICT; Final answer: ODDS and ENDS. Students may use strategies, such as trying different permutations

(continued)

7. Scrambled Word Puzzles

or trying to solve the puzzle first, to narrow down the number of possible arrangements of the letters.

Extension Activity

Have students search the Internet for sites concerning scrambled word puzzles. Some such sites have interactive programs that supply answers for scrambled words. Ask students to speculate on how such programs might work.

7. Scrambled Word Puzzles

Many daily newspapers feature a puzzle in which four or five words are scrambled. You figure out what the words are, then use letters found in certain positions of these words to decode a joke or clever phrase. Use your knowledge of probability to solve this type of puzzle.

1. How many ways is it possible to arrange the 4 letters *A, P, S,* and *T*? Write all the possibilities below. How many of the possibilities are actual words?

    number of possible arrangements _____

    number of actual words _____

2. How many ways is it possible to arrange

    a. 5 distinct letters?

    b. 6 distinct letters?

    c. the letters in the word *GEESE*?

    d. two groups of 3 distinct letters?

3. Complete the puzzle below. Describe any strategies that you used to unscramble the words.

Probability Puzzler
Unscramble these four words, putting one letter in each square, to form four words related to probability.

N A R M O D

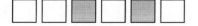

S P E S M L A

D C E E U D

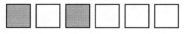

C I T R D E P

Now arrange the letters in the gray boxes above to answer the following riddle.

What did Luis find when he turned to the back of his probability textbook?

☐☐☐☐ AND ☐☐☐☐

Real-Life Math: Probability

8. Trials: Single Coin Toss

Context

sports

Math Topic

probability distributions

Overview

There is no better way to understand probability than to do simple experiments. Perhaps the easiest way to generate random events is to toss a coin.

Use Activity 9 if you wish to study the binomial model on which these trials are based.

Objectives

Students will be able to:

- conduct and record the results of experimental trials

- make a graph to analyze the results

Materials

- one copy of the Activity 8 handout for each pair of students

- one coin for each pair of students

- graph paper

- calculator

Teaching Notes

Students should pair off. Within each pair, one student will toss or spin the coin to produce good random trials. The other will record the results. Students should switch jobs halfway through the trials. It is recommended that the person tossing the coin catch it rather than let it bounce on the floor.

Students will record 20 trials of 10 tosses each. The exact sequence of heads and tails in each trial should be written on the sheet. For each trial, the students should count and record the number of times the outcome was heads.

Students can then complete a histogram showing the frequency with which heads occurred 0 times, 1 time, 2 times, and so forth, up to 10 times in the trials.

If you decide to do Activity 9, make sure that students save their data from this activity.

Answers

1–2. Results will vary. The theoretical distribution of probability is

0H, $1/2^{10}$

1H, $10/2^{10}$

2H, $44/2^{10}$

3H, $117/2^{10}$

4H, $205/2^{10}$

(continued)

8. Trials: Single Coin Toss

5H, $246/2^{10}$

6H, $205/2^{10}$

7H, $117/2^{10}$

8H, $44/2^{10}$

9H, $10/2^{10}$

10H, $1/2^{10}$

For 20 iterations of the 10 flips, occurrences of 0H, 1H, 9H, and 10H will be rare, with total probability under 5% that any of these will happen. However, many students will experience one or perhaps two of these rare events within their 20 iterations. The rest of the theoretical distribution is 2H, 1 occurrence; 3H, 2 occurrences; 4H, 4 occurrences; 5H, 5 occurrences; 6H, 4 occurrences; 7H, 2 occurrences; 8H, 1 occurrence.

Extension Activity

Have students recreate this activity with a die, investigating the probability of any one of the 6 sides being rolled. One partner will roll the die, and the other will record the results. They should record 20 trials of 10 rolls each. After all the data are collected, ask students to make a histogram showing the frequency with which their selected number (1–6) occurred.

8. Trials: Single Coin Toss

You know there is a 50–50 chance that your team will win the coin toss at the beginning of a football game. But what is your team's probability of coming out ahead (or behind) in winning the toss during the course of a 10-game season? Find out in this simulation.

1. One partner will toss or spin a coin to produce good random trials. The other will record the results on this sheet. Switch jobs halfway through the trials. It is recommended that the person tossing the coin catch it rather than let it bounce on the floor. Record 20 trials of 10 tosses each. The exact sequence of heads and tails should be written in the table under "Outcomes." For each trial, record the number of times the outcome was heads out of the 10 tosses.

Trial	No. of Heads	Outcomes	Trial	No. of Heads	Outcomes
1			11		
2			12		
3			13		
4			14		
5			15		
6			16		
7			17		
8			18		
9			19		
10			20		

2. After all data are obtained, make a histogram on a sheet of graph paper showing the frequency with which heads occurred 0 times, 1 time, 2 times, and so forth, up to 10 times in the trials.

9. Analysis: Single Coin Toss

Context

simple experiment

Math Topic

binomial model

Overview

In this activity, students use their data from Activity 8 to study the binomial model. Any independent trials built on events that have two possible outcomes, success or failure, fall under the binomial model. In this activity, students study the single coin toss, where success is defined as an outcome of heads and failure as an outcome of tails.

Objectives

Students will be able to:

- understand the binomial distribution

- compare experiment and theory for 10 coin tosses

Materials

- one copy of the Activity 9 handout for each student

- data from Activity 8

- calculator

Teaching Notes

To determine the likelihood of getting a certain number of heads out of 10 tosses of a coin, it is necessary to understand a mathematical model of this situation: the binomial model. This activity lets students calculate the binomial probabilities for their data and compare the calculated probabilities to their actual results.

The term *binomial* follows from the expression $(p + q)^n = 1$. When this expression is expanded into all of its binomial terms, the terms, each of which represents a certain number of successes and failures, all add up to 1. The general expression for an individual binomial probability (P) is:

$$P_{n,m} = \binom{n}{m} p^m q^{(n - m)}$$

In this expression, n is the number of trials, m is the number of trials with successes, p is the probability of success, and q is the probability of failure $(1 - p)$.

It may be easier to teach students how to use the binomial model for a specific case. Say you want to know, before tossing a coin 10 times, the probability of getting 7 heads. First, you need to determine how many ways there are to "choose" 7 events (i.e., get heads 7 times) out of 10 events (i.e., 10 tosses), because any combination of 7 events being heads would result in a total of 7. From our discussion on combinations, we know there are $(10 \times 9 \times 8 \times 7 \times 6 \times 5 \times 4)/7!$, or 120, such combinations.

(continued)

9. Analysis: Single Coin Toss

Then you must multiply this number by the probability of any individual series of 10 tosses coming out with 7 heads. The probability of heads (success) is $\frac{1}{2}$, and the probability of tails (failure) is also $\frac{1}{2}$. Because all 10 events are independent, this probability is $(\frac{1}{2} \times \frac{1}{2} \times \frac{1}{2} \times \frac{1}{2} \times \frac{1}{2} \times \frac{1}{2} \times \frac{1}{2}) \times (\frac{1}{2} \times \frac{1}{2} \times \frac{1}{2}) = (\frac{1}{2})^7 \times (\frac{1}{2})^3 = (\frac{1}{2})^{10} = \frac{1}{1024}$.

The total probability of 7 out of 10 heads is $120 \times \frac{1}{1024} = 0.117$. Students should be able to calculate the remaining values.

Answers

1. Compare students' results with this theoretical probability distribution.

    0H, 0.001

    1H, 0.010

    2H, 0.044

    3H, 0.117

    4H, 0.205

    5H, 0.246

    6H, 0.205

    7H, 0.117

    8H, 0.044

    9H, 0.010

    10H, 0.001

    See the answers for Activity 8 for additional comments.

2. Rarely will actual results exactly match the theoretical distribution. This would be a good point for a discussion about the nature of events with random properties and the meaning of mathematical probability. The most important lesson here is that the mathematical probability of $\frac{1}{2}$ reflects a distribution of results with a mode at 5 of 10 tosses coming up heads.

Extension Activity

Have students recalculate the table assuming your trials involve two coins and that success is defined as an outcome of both heads. Run actual trials if desired. The probability of success is then $\frac{1}{4}$. Students will apply the binomial model for $p = \frac{1}{4}$ and $q = \frac{3}{4}$.

9. Analysis: Single Coin Toss

1. Complete the table showing the probability of each of the possible outcomes of a series of 10 coin tosses. Use the data you collected in Activity 8.

    Number of trials from Activity 8: _____

No. of heads	No. of combinations	Probability of individual outcome	Total probability	No. of heads from actual trials	Percent of heads from actual trials
0					
1					
2					
3					
4					
5					
6					
7					
8					
9					
10					

2. What have you learned about the meaning of a mathematical probability of $\frac{1}{2}$ from observing and recording many actual events with this probability?

     Real-Life Math: Probability

10. Two Dice: When Will Doubles Occur?

Context

games

Math Topic

waiting times

Overview

Activity 10, Activity 11, and Activity 12 are designed to work together. You may do Activity 10 and Activity 12, Activity 11 and Activity 12, or all three activities.

Dice provide an excellent tool for studying the properties of independent events. When playing games involving dice or other random devices, many people feel that they become "due" for a particular outcome. In reality, for each repetition, the event has exactly the same probability.

Objectives

Students will be able to:

- understand that when independent events are repeated, outcomes do not depend on previous results

- measure waiting times for discrete, independent events

Materials

- one copy of the Activity 10 handout for each student

- two standard six-sided dice for each student

- calculator

Teaching Notes

First, students should create a chart showing all the possible outcomes for rolling a pair of dice. Each outcome has a probability of 1 of 36. Students should be able to conclude that 6 of 36 outcomes are doubles. Therefore, the probability of doubles is $^6/_{36} = ^1/_6$.

Next, each student should do his or her own trials. The results are easy to record because there is only one value to write down for each trial—the number of rolls that it took to get doubles. Sixty trials should take about 30 minutes.

Students are directed to wait for instructions after the trials. You may either proceed to Activity 11 to collect data concerning the sum on a pair of dice or go directly to Activity 12 to analyze the data.

Make sure students keep their data sheets for Activity 12. To get a larger sample size, combine everyone's data from this activity and then use Activity 12 as a whole-class activity to analyze the data.

(continued)

10. Two Dice: When Will Doubles Occur?

Answers

1. a. 36

   b. 6

   c. 1/6

2. Student results will vary and will be analyzed in Activity 12.

Extension Activities

- Students may study games of chance that involve dice and report on the probabilities of various outcomes in the games.

- Students may expand their study to games with three or more dice.

10. Two Dice: When Will Doubles Occur?

You love to play Monopoly®, but you hate to land in jail. You need to roll doubles to get out. How long might you have to wait in jail? Complete this activity to find out.

1. a. How many total outcomes are in the sample space for two dice?

    b. How many outcomes are doubles?

    c. What is the probability of rolling doubles on a single roll?

2. Conduct a series of trials in which you roll two dice repeatedly. Use the table below to record your results. A trial will consist of counting the number of rolls until doubles occurs. Record the number of rolls you count for each trial in the table. Start the next trial and count the number of rolls until doubles occurs again. If you don't get doubles after the twentieth roll, call the count 20 and start a new trial. Perform 30 to 60 trials, depending on the amount of time available.

Trial number	Number of Rolls	Trial number	Number of Rolls	Trial number	Number of Rolls	Trial number	Number of Rolls	Trial number	Number of Rolls	Trial number	Number of Rolls
1		11		21		31		41		51	
2		12		22		32		42		52	
3		13		23		33		43		53	
4		14		24		34		44		54	
5		15		25		35		45		55	
6		16		26		36		46		56	
7		17		27		37		47		57	
8		18		28		38		48		58	
9		19		29		39		49		59	
10		20		30		40		50		60	

Follow your teacher's instructions to report and analyze your data.

11. Two Dice: Sums

Context

games

Math Topic

waiting times

Overview

Activity 10, Activity 11, and Activity 12 are designed to work together. You may do Activity 10 and Activity 12, Activity 11 and Activity 12, or all three activities. These activities allow students to study the probabilities of the possible outcomes of rolling a pair of dice.

Objectives

Students will be able to:

- understand that when independent events are repeated, outcomes do not depend on previous results

- measure waiting times for discrete, independent events

Materials

- one copy of the Activity 11 handout for each student

- two standard six-sided dice for each student

- calculator

Teaching Notes

Students should first complete the chart showing the sums of all the possible outcomes for rolling a pair of dice. Each outcome has a probability of 1 in 36. Students should be able to add up the number of outcomes that represents each sum.

Next, each student should do his or her own trials. The results are easy to record because there is only one value to write down for each trial: the number of rolls it takes to get a chosen sum. Sixty trials should take about 30 minutes, but you may adjust the number of trials to accommodate the allotted time.

Students are directed to wait for instructions after the trials. You may either go to Activity 10 to collect more data concerning waiting for doubles or go directly to Activity 12 to analyze the data.

Make sure students keep their data sheets for Activity 12. To get a larger sample size, combine everyone's data from this activity and use Activity 12 as a whole-class activity to analyze the data.

(continued)

teacher's page

11. Two Dice: Sums

Answers

1. **Possible Outcomes**

Die 2 \ Die 1	1	2	3	4	5	6
1	2	3	4	5	6	7
2	3	4	5	6	7	8
3	4	5	6	7	8	9
4	5	6	7	8	9	10
5	6	7	8	9	10	11
6	7	8	9	10	11	12

Sum	Outcomes	Probability
2	1	1/36
3	2	1/18
4	3	1/12
5	4	1/9
6	5	5/36
7	6	1/6
8	5	5/36
9	4	1/9
10	3	1/12
11	2	1/18
12	1	1/36

2. Student results will vary and will be analyzed in Activity 12.

Extension Activities

- Students may study games of chance that involve dice and report on the probabilities of various outcomes in the games.
- Students may expand their study to games with three or more dice.

Name ______________________________ Date ______________________

11. Two Dice: Sums

1. Complete the tables below showing the sums for all possible outcomes for rolling two dice. The first few possible outcomes are already recorded. Then record in each table the total number of outcomes and probability of rolling each possible sum.

Possible Outcomes

Die 1

Die 2	1	2	3	4	5	6
1	2	3				
2	3					
3						
4						
5						
6						

Sum	Outcomes	Probability
2		
3		
4		
5		
6		
7		
8		
9		
10		
11		
12		

2. Choose a sum. Conduct a series of trials in which you roll two dice repeatedly. Use the table below to record your results. A trial will consist of counting the number of rolls until your sum occurs. Record the number of rolls for each trial in the table. Start the next trial and count the number of rolls until the sum occurs again. If you don't get your sum after the twentieth roll, call the count 20 and start a new trial. Perform 30 to 60 trials (according to your teacher's instructions).

Trial	Rolls	Trial	Rolls	Trial	Rolls	Trial	Rolls	Trial	Rolls	Trial	Rolls
1		11		21		31		41		51	
2		12		22		32		42		52	
3		13		23		33		43		53	
4		14		24		34		44		54	
5		15		25		35		45		55	
6		16		26		36		46		56	
7		17		27		37		47		57	
8		18		28		38		48		58	
9		19		29		39		49		59	
10		20		30		40		50		60	

Follow your teacher's instructions to report and analyze your data.

12. Analysis: Doubles or Sums

Context

games

Math Topic

waiting times

Overview

This activity is a continuation of Activity 10 and Activity 11. It provides an opportunity to analyze data from one or both of these activities and to take a deeper look at waiting-time distributions.

Objectives

Students will be able to:

- analyze data for waiting-time trials

- show graphically the frequency distribution of a waiting-time experiment

Materials

- one copy of the Activity 12 handout for each student

- graph paper

- waiting-time data from Activity 10 or Activity 11

- calculator

Teaching Notes

All students must have waiting-time data from Activity 10, Activity 11, or both. Students will work with calculators to tabulate and analyze their data. You may want to do this activity as a class exercise by combining everyone's data into one set of trials. If you do this, reproduce the summary table and distribution histogram on the board or the overhead projector.

Answers

1. a. Students should enter *doubles* or *sums* along with the sum (2–12) that they were trying to achieve.

   b. For doubles, the theoretical probability is $1/6$; for a sum:

      of 2, $1/36$

      of 3, $1/18$

      of 4, $1/12$

      of 5, $1/9$

      of 6, $5/36$

      of 7, $1/6$

      of 8, $5/36$

      of 9, $1/9$

      of 10, $1/12$

      of 11, $1/18$

      of 12, $1/36$

(continued)

     Real-Life Math: Probability

12. Analysis: Doubles or Sums

2. a. The table will reflect student results. This sample shows the theoretical values for a probability of 1/6 and 60 total repetitions. Trials seeking doubles or a sum of 7 have this probability. The table would look different for different probabilities. See Activity 10 or 11 for a discussion about how to find these values. Student results will rarely match exactly the theoretical results.

   b. The cumulative percentage of doubles (or a sum of 7) occurring exceeds 50% by the fourth roll. In other words, in 50% of the trials, four or fewer rolls of the dice were necessary to achieve the desired outcome. Here are cumulative percentage values for the other sums: 2 or 12, more than 20 rolls; 3 or 11, thirteen rolls; 4 or 10, eight rolls; 5 or 9, six rolls; 6 or 8, five rolls.

3. Student results will vary. They will rarely match theoretical values, but in most cases, the histogram will show maximum values for one roll and steadily decreasing values for more rolls. For outcomes with lower probability, the histogram will be flatter.

Extension Activity

Have students set up computer spreadsheets to analyze waiting times.

Rolls	No. of Trials	% of Trials	Cuml. %
1	10	16.7%	16.7%
2	8	13.3%	30.9%
3	7	11.7%	41.7%
4	6	10.0%	51.8%
5	5	8.3%	60.0%
6	4	6.7%	66.7%
7	3	5.0%	71.7%
8	3	5.0%	76.7%
9	2	3.3%	80.0%
10	2	3.3%	83.3%
11	2	3.3%	86.6%
12	1	1.7%	88.3%
13	1	1.7%	90.0%
14	1	1.7%	91.7%
15	1	1.7%	93.4%
16	1	1.7%	95.1%
17	1	1.7%	96.8%
18	0	0.0%	96.8%
19	0	0.0%	96.8%
20*	2	3.3%	100.0%

* The frequency of 20 rolls is elevated because it includes all values equal to or greater than 20.

12. Analysis: Doubles or Sums

1. This sheet is to be used to analyze your results from the waiting-times trials for doubles or sums. Please indicate whether this sheet is for doubles or sums trials. If it is for sums, indicate which sum you desired in your trials.

    a. This sheet is for waiting-time trials for: _____

    b. The theoretical probability for this outcome in a single trial is: _____

    c. How many total trials were performed? _____

2. a. In the table below, tabulate how many trials required the indicated number of rolls to reach the desired outcome. Calculate the percentage of trials for each number of rolls and the cumulative percentage of trials.

Rolls	No. of Trials	% of trials	Cuml. %	Rolls	No. of Trials	% of trials	Cuml. %
1				11			
2				12			
3				13			
4				14			
5				15			
6				16			
7				18			
8				18			
9				19			
10				20			

    b. How many rolls are necessary before the probability of occurrence of the desired outcome is at least 50%?

3. Use a sheet of graph paper to make a histogram showing how often each number of rolls was required to get doubles or your selected sum from the data in your trials.

13. Password Possibilities

Context

computers

Math Topic

multiplication and exponents

Overview

Many types of electronic security depend on passwords. But there are a finite number of possibilities for passwords of finite length. This activity focuses on using multiplication and exponents to count possibilities.

Objectives

Students will be able to:

- count possible passwords using multiplication and exponents

Materials

- one copy of the Activity 13 handout for each student

- calculator

Teaching Notes

You may need to review the basics of exponents. Students should know the meaning of mathematical statements such as 2^{10} or 26^5. They will need their calculators and need to know how to use them to find the values of expressions with exponents.

The most interesting question is 1d, which illustrates how many binary digits would be required in a code so that even the fastest computers could not guess every possible combination in a reasonable amount of time. You many want to convert answers in seconds to days or years.

Answers

1. a. 2 or 2^1, Heads (H) or Tails (T)

    b. 4 or 2^2, HH, HT, TH, and TT

    c. 2^3, 2^4, 2^8, 2^{16}, and 2^{48}

    d. Any answer 30 or lower is bad because the computer could check all the possibilities in 1 second or less. Even 40 coins would be inadequate because only 2^{10} or 1024 seconds would be required. Numbers larger than 100 would be adequate. The computer would require more than 1013 years to check all the possibilities.

2. There are 12^3 or 1728 possible combinations. The thief will test 50% in $(1728 \times 0.50)/100$ or 8.64 minutes.

3. a. $26^3 = 17{,}576$

    b. $26^5 = 11{,}881{,}376$

    c. $26^6 = 308{,}915{,}776$

Extension Activities

- Students could research the use of encryption in computer security.

- Ask students why computer service providers recommend passwords that are at least eight characters long, using both letters and numbers.

13. Password Possibilities

Combination locks are everywhere. You probably have one on a locker. But how secure is such a lock? And what about computer security? What is the probability that someone could hack into your computer or e-mail account? It could happen unless you are careful about how you choose your password.

1.   Let's say you want to build a security code based on coin tosses. Your code will be based on the outcome of one or more coins tossed at the same time. To break the code, an intruder (perhaps using a computer) must guess the exact pattern of heads and tails in your coin toss.

   a.   How many ways can a single tossed coin land? Write down the possibilities and write this number of possibilities as a power of 2.

   b.   How many ways can two tossed coins land? Write down the possibilities and write this number of possibilities as a power of 2.

   c.   How many possible ways can 3, 4, 8, 16, and 48 coins land? Write these numbers only as powers of 2.

   d.   A certain computer can check 2^{30} patterns of heads and tails per second. How many coins would be needed to make a code that is safe from this computer?

(continued)

13. Password Possibilities

2. You have an old-style combination lock with three wheels, each containing 12 numbers. Exactly one combination of the three numbers must be selected to open the lock. A thief trying to open the lock checks 100 combinations per minute, at random. In how many minutes will there be a 50% probability that the thief will open the lock?

3. Internet services usually require passwords. Assume that you must choose a password using only the standard 26 letters (no symbols or numbers). How many different passwords can you choose with

    a. 3 letters?

    b. 5 letters?

    c. 6 letters?

14. Setting Up a Basketball Tournament

Context

sports

Math Topic

combinations

Overview

When setting up a basketball tournament, organizers want the two best teams to have the highest probability of meeting in the final game. Past records are the only data available to help design the brackets.

Objectives

Students will be able to:

- study the number of possible combinations in an 8-team tournament

- propose tournament pairings with the best chance of having the best teams in the final game

Materials

- one copy of the Activity 14 handout for each student

Teaching Notes

The first question involves just figuring out how many ways an 8-team tournament could be set up. The result is surprisingly large. The second question asks students to eyeball the win–loss records and set up a tournament. Then they guess the probability that the final game would be a disappointment.

Answers

1. a. 28

   b. 15

   c. 6

   d. 1

   e. $28 \times 15 \times 6 \times 1 = 2520$

2. Student results will vary. The only requirement is that the Wolves and Bears meet in the final game.

3. This question requires students to guess the probabilities of the best teams not winning their first game, or winning the first then losing the second. These probabilities would be added to get the result. For example, say the Wolves are paired with the Sharks in the first game and the Wolves have a probability of only 0.05 of losing to the Sharks. They have a much higher probability of losing to any second-round opponent, say 0.25. Then the probability of the Wolves losing before the final game is $(0.05) + (0.95)(0.25) = 0.29$. Any reasoning along these lines is acceptable. There is a pretty good chance that one or both of the two best teams won't make the finals.

Extension Activity

Have students study pairings from recent sports tournaments and decide if the best teams did meet in the finals.

14. Setting Up a Basketball Tournament

You are the director of a basketball tournament. Eight teams are invited. You'd like to have the two best teams meet in the final game for the championship. Below is a table with the 20-game season records of all of the invited teams. (Some teams played teams not listed.)

	Wolves	Bears	Tigers	Lions	Eagles	Hawks	Bulldogs	Sharks
Won-Lost	19–1	18–2	15–5	12–8	8–12	7–13	6–14	3–17
Percent	0.95	0.90	0.75	0.60	0.40	0.35	0.30	0.15

1. a. How many different pairs of the 8 teams are possible for games in the first round?

    b. After one pair of teams is matched, how many different pairs from the remaining 6 teams are possible for the second game?

    c. How many different pairs from the remaining 4 teams are possible for the third game?

    d. How many pairs are left for the fourth game?

    e. What is the total number of possible ways to set up pairs for the tournament?

(continued)

14. Setting Up a Basketball Tournament

2. Set up the tournament using your best judgment about how to arrange pairings so that the two best teams will meet in the final game. Use the win–loss record on the previous page to guide you.

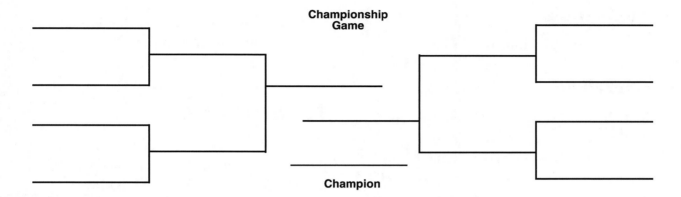

Championship Game

Champion

3. Guess the probability from your setup that the Wolves will be knocked out before the final round.

15. Winning a Best-of-7-Game Series

Context

sports

Math Topic

combinations and multiplication rule

Overview

Mathematical models are often used to describe the pattern of data from an experiment. Most of the examples in this book follow simple models and have well-defined sample spaces. For example, tossing a coin has two possible outcomes: heads or tails. A series of such tosses can be shown to follow the binomial model studied in Activities 8–12.

However, in some situations the underlying probabilities can only be estimated from past experience. Sports is a good example. Two teams that meet in a best-of-7-game playoff series often have played many times in the past. Given this experience, what is the probability that the underdog will win the series?

Objectives

Students will be able to:

- use combinations to find the number of ways a sports team can win 4 out of 7 games in a playoff series

- conduct simulations of playoff series to see how often the underdog wins

Materials

- one copy of the Activity 15 handout for each student

- one random number table for each student (see Appendix)

- calculator

Teaching Notes

These questions lead students through the process of using the binomial model for independent trials. This model allows calculation of the probability of m successes in n tries as long as the underlying probability is known. This probability may or may not be valid for sports teams in an actual playoff. But a long-term winning percentage of only 30% against an opponent makes the team discussed here a clear underdog. This could be a matter for class discussion.

Students are given hints about how to calculate outcomes and probabilities. They must apply what they have learned about combinations and the multiplication rule. You probably will need to help them. Here's a sample for a 7-game series in which the underdog wins after game 7:

Number of ways team can win in 7 games

$$= \binom{6}{3} = \frac{6 \times 5 \times 4}{3 \times 2 \times 1} = 20$$

Probability the underdog wins a 7-game series:

$$\binom{6}{3}(0.3)^4 (0.7)^3 = 20 \times 0.002778 = 0.05557$$

(continued)

15. Winning a Best-of-7-Game Series

This activity is a good one to introduce the random number table. The simulation is very simple: Look up random numbers one at a time until one team wins four, then record the series winner and repeat. The underdog wins on 01 through 30. The opponent wins on 31 through 00 (00 represents 100). Make sure students use a new sequence of numbers for every series played.

2. Student results will vary. If there is enough time, a series-winning percentage of 13% will emerge with enough trials.

Extension Activity

Use sports record books for any sport with a best-of-7-game championship series to see how often the underdog wins.

Answers

1. a. 1

   b. 4

   c. 20

   d. $0.3 \times 0.3 \times 0.3 \times 0.3$ or 0.81%

   e. $4 \times 0.3 \times 0.3 \times 0.3 \times 0.3 \times 0.7$ or 2.3%

   f. $10 \times 0.3 \times 0.3 \times 0.3 \times 0.3 \times 0.7 \times 0.7$ or 4.0%

   g. $20 \times 0.3 \times 0.3 \times 0.3 \times 0.3 \times 0.7 \times 0.7 \times 0.7$ or 5.6%

   h. 0.81% + 2.3% + 4.0% + 5.6% or 12.7%

15. Winning a Best-of-7-Game Series

You're a fan of a team that goes into a best-of-7-game championship series with a dreaded opponent. But there's a big problem—your team has won only 30% and has lost 70% of its many games played against this opponent over the last couple of years. Your team is the clear underdog. Is there any hope? Yes, but they're going to need some luck.

1. a. How many outcomes are there in which your team wins a 4-game sweep?

   b. How many outcomes are there in which your team wins with the fifth game? Remember, 3 of your team's 4 wins can come in any 3 of the first 4 games.

   c. How many outcomes are there in which your team wins with the seventh game? Remember, 3 of your team's 4 wins can come in any 3 of the first 6 games.

   d. What is the probability of your team sweeping 4 games? (*Hint:* Use the multiplication rule and their long-term winning percentage as the probability.)

   e. What is the probability of your team winning in exactly 5 games? (*Hint:* Multiply the total number of outcomes where your team wins after 5 games by the probability of 4 wins and the probability of 1 loss.)

   f. What is the probability of your team winning in exactly 6 games?

   g. What is the probability of your team winning in exactly 7 games?

   h. What is the probability of your team winning the series?

(continued)

15. Winning a Best-of-7-Game Series

2. A random number table gives you a way to test probability models. Your table has 1000 two-digit numbers selected completely at random. You can use it to quickly simulate many 7-game series to see how often your team wins.

Choose numbers from the table one at a time in sequence. If any number from 01 to 30 comes up, your team wins a game. If 31 through 00 comes up, the opponent wins. As soon as one team wins a fourth game (this will happen in a maximum of 7 games), the series is over. Record the winner below. Play as many series this way as you can. In what percentage of the series does your underdog team win?

Your team wins series:	Opponent wins series:
Total for your team:	Total for opponent:
Percentage of series won:	Percentage of series won:

16. What Is a Fair Price?

Context

recreation

Math Topic

expected value

Overview

This activity examines expected value, an average that arises when dealing with many repetitions of probabilistic situations in which different outcomes result in different returns. Expected value is very important in gaming situations because it reveals how much money will change hands in the long run, after many repetitions of the game. Furthermore, expected value can tell a player if the amount paid to play is a fair price.

Objectives

Students will be able to:

- demonstrate calculation of expected value

- determine the long-term average return from a game

Materials

- one copy of the Activity 16 handout for each student

- two standard six-sided dice for each pair of students

- calculator

Teaching Notes

What is a fair game? Try to get students to reason that a game of chance would be absolutely fair if no one gained or lost money over many plays. But no lottery or any other game of chance is absolutely fair. Games are always set up so that the provider (e.g., the state, the casino operator) makes money in the long run.

The exercises in this activity examine the following concept from probability theory:

$$\text{Expected value} = \sum_{i=1}^{n} f_i p_i$$

In this equation, f_i is equal to the prize for outcome i, p_i is equal to the probability of outcome i, and n is the total number of outcomes. Emphasize that this will yield a player's average winnings per play over a large number of repetitions. For state lotteries, this value usually is set to be about $0.50 per $1 bet. Thus, the regular lottery player can expect to lose half of his or her money in the long run. For lotto games where the large prize is quite rare, the experience of most players will be worse yet.

(continued)

16. What Is a Fair Price?

Answers

1. a. The expected value is $2(1/4) + $1(1/2) = $1 average player winnings per play. So $1 is the break-even price over the long term, and a price of $2 will result in 50–50 split.

    b. The expected value is $6($1/_6$) + $18($1/_{18}$) = $2 average player winnings per play. So $2 is the break-even price over the long term, and a price of $4 will result in 50–50 split.

    c. Students first will have to calculate the total number of possible outcomes. See if they can reason this out before helping them. The total number of possible outcomes is (12 • 11 • 10) / 3! = 220. Students are then given the number of ways that three, two, or one number can match out of the 220 possible outcomes. The expected value is $100($1/_{220}$) + $5($27/_{220}$) + $1($108/_{220}$) = $1.56 average player winnings per play. So $1.56 is the break-even price over the long term, and a price of $3.12 will result in a 50–50 split.

2. Students should assume that each roll costs the player $4, as calculated in 1b. Then for 60 rolls, the player will spend $240. To determine the amount won by the player, multiply the number of times the player rolls 11 by $18 and the number of times the player rolls 7 by $6 and add these amounts. To determine the amount gained by charity, subtract the amount won by the player from the $240 spent by the player.

Extension Activity

Have students research the expected winnings per dollar wagered for various types of gambling—state lotteries, casinos, bingo, and so forth.

16. What Is a Fair Price?

A charity has set up some games to raise money. It is your job to calculate the break-even ticket price and the price that should be charged so that the charity will keep 50% of the money collected and pay out 50% in prizes. Assume that enough tickets to cover the prizes are always sold.

1. a. The game is a coin toss where players toss a coin twice. Players win $2 if they get heads on both tosses, $1 for one head, and $0 for both tails.

    break-even price _____

    50–50 price _____

    b. The game is rolling two dice. Players win $6 if they roll a sum of 7 and $18 if they roll a sum of 11.

    break-even price _____

    50–50 price _____

    c. The game is a lotto game where players choose 3 numbers out of 12. The game operator chooses the 3 winning numbers. If all 3 picks match (1 possible way), the player wins $100. If 2 of 3 match (27 possible ways), the player wins $5. If 1 of 3 match (108 possible ways), the player wins $1. (There is no prize for no matches.)

    break-even price _____

    50–50 price _____

2. Run a simulation of part 1b above. Work with a partner. One of you should act as the player and roll the dice. The other should act as the charity and record the data. Imagine that the player pays the price calculated for a 50–50 split before each roll. How much has the player spent and won back and how much has the charity gained after 60 rolls of the dice?

    spent by player _____

    won by player _____

    gained by charity _____

17. Number of Boys and Girls in a Family

Context

families

Math Topics

geometric distribution and averages

Overview

This topic is a further application of probability principles learned in Activities 8–14. Assuming that for a live birth, the probability of a baby of either sex is $\frac{1}{2}$, the situation is similar to tossing a coin. In this activity, students will calculate the probabilities of various family compositions of boys and girls, then conduct trials concerning waiting times for a family to have at least one boy and one girl. The concept of an average (or expected value) is introduced.

Objectives

Students will be able to:

- calculate the probabilities of family composition for families of up to 6 children

- determine the average of a set of trial data concerning family composition

Materials

- one copy of the Activity 17 handout for each pair of students

- one coin or a random number table for each pair of students (see Appendix for random number tables)

- calculator

Teaching Notes

The assumption that the probability of a new baby being a girl (or a boy) is exactly $\frac{1}{2}$ is fundamental to this activity. You may want to discuss this as a class, or as an extension, you might ask students to research actual statistics. There are many interesting possibilities here, especially if sex distribution is considered in countries other than the United States.

This probability is a geometric distribution. The distribution starts at two children (because there must be at least two to have a boy and a girl). So the probability (P) of n children ($n > 1$) is thus:

$$P(n) = \left(\frac{1}{2}\right)^{(n-1)}$$

The activity considers parents who plan to stop having children when they have at least one boy and one girl. The trials may be conducted with a coin or a random number table.

After the trials are completed, students are asked to find the average number of children in a family with the given conditions. Make sure students know how to find an average (or mean). The average is equal to the sum of the counts from all trials divided by the number of trials.

(continued)

17. Number of Boys and Girls in a Family

Answers

1. Student results for the table will vary.

2. Students could count the number of trials in which the number of kids is greater than 4, then divide by the total number of trials. Theoretically, 7/8 of the trials will yield 2, 3, or 4 children, and 1/8 will have higher values.

3. The theoretical average is 3 children.

Extension Activities

- Have students research factors that affect family size and composition.

- Have students develop the following probability table for family composition for families with 1 to 6 children using the binomial model. Note that the total probability for each line sums to 1.

Children in family	0 girls	1 girl	2 girls	3 girls	4 girls	5 girls	6 girls
1	$\frac{1}{2}$	$\frac{1}{2}$					
2	$\frac{1}{4}$	$\frac{1}{2}$	$\frac{1}{4}$				
3	$\frac{1}{8}$	$\frac{3}{8}$	$\frac{3}{8}$	$\frac{1}{8}$			
4	$\frac{1}{16}$	$\frac{1}{4}$	$\frac{3}{8}$	$\frac{1}{4}$	$\frac{1}{16}$		
5	$\frac{1}{32}$	$\frac{5}{32}$	$\frac{5}{16}$	$\frac{5}{16}$	$\frac{5}{32}$	$\frac{1}{32}$	
6	$\frac{1}{8}$	$\frac{3}{32}$	$\frac{15}{64}$	$\frac{20}{64}$	$\frac{15}{64}$	$\frac{3}{32}$	$\frac{1}{64}$

Name _____   Date _____

17. Number of Boys and Girls in a Family

A young couple is discussing how many children they want to have. The man says, "Let's have as many children as necessary until we have at least one girl and one boy, then stop."

The woman says, "I'll go as high as four. After that, you can bear the children yourself!"

What is the probability that they can achieve their desired family size and composition?

1. To answer the question above, conduct trials with a random number table. A trial will consist of counting random numbers until there is at least one even number (representing a girl) and one odd number (representing a boy) in the series. Record your count until an even and an odd number are obtained for each trial in the table below. This count will represent the total number of children in the family. The minimum count for each trial is 2. (Alternately, you may toss a coin to conduct the trials, counting the number of flips until you have one head and one tail.)

Trial	Kids	Trial	Kids	Trial	Kids	Trial	Kids	Trial	Kids	Trial	Kids
1		11		21		31		41		51	
2		12		22		32		42		52	
3		13		23		33		43		53	
4		14		24		34		44		54	
5		15		25		35		45		55	
6		16		26		36		46		56	
7		17		27		37		47		57	
8		18		28		38		48		58	
9		19		29		39		49		59	
10		20		30		40		50		60	

2. What is the probability that the result is 4 children or fewer?

3. What would be the average number of children for a large number of families where parents have children until they have at least 1 boy and 1 girl?

18. Chance of a Hitting Streak

Context

sports

Math Topic

waiting time

Overview

In baseball and softball, individual performances are the key to team success. Players are rated by many statistics, the most important of which is the batting average. This statistic is really a probability: the number of times out of 1000 that a player gets to base safely by hitting the ball, and no defensive player makes an error.

It's common for players to "streak," that is, to experience hot periods with many hits or cold periods with few hits. In this activity, students explore the probabilities of players' hitting streaks given their batting averages.

Objectives

Students will be able to:

- calculate the probabilities of a player with a particular batting average either getting a hit or not getting a hit in a game

- determine the average hitting streak from a set of trial data

Materials

- one copy of the Activity 18 handout for each pair of students

- one random number table for each pair of students (see Appendix)

- calculator

Teaching Notes

This activity is similar to Activity 17 in that it considers waiting times. Students must calculate the probabilities of a player with a given batting average getting at least 1 hit out of 4 at-bats (consider 4 at-bats to be a "game") and getting no hits out of 4 at-bats. Then they run trials to determine the probability of having several games in a row with at least 1 hit (a "hitting streak").

To run the trials, students use the probability for getting a hit in a game that they have calculated in 1c. They round this probability to two significant digits. Their success is achieved when a number from 01 through the two-digit value appears on the random number table. Students should be aware that probabilities add to 1, so $P(A) = 1 - P(\text{not } A)$.

(continued)

 Real-Life Math: Probability

18. Chance of a Hitting Streak

Answers

1. It is easier to do part b first. The probability of no hits is $(1 - \text{batting average})^4$. So the answers are as follows:

   a. probability of at least 1 hit =
   $(1 - \text{probability of no hits}) = 1 - (1 - \text{batting average})^4 = 1 - (0.68)^4 = 0.786$

   b. $(1 - \text{batting average})^4 = 0.214$

2. Answers will vary.

3. Answers will vary with respect to the average chosen and the individual trials.

Extension Activities

- Use the newspaper to track hitting streaks of major league baseball players.

- Students may want to track the streaks of players on a school or other local team.

18. Chance of a Hitting Streak

Hitting streaks (consecutive games getting a hit) are one of the most exciting individual performance statistics in baseball and softball.

1. a. If a baseball or softball player has a batting average of .320, what is the probability that he or she will get at least 1 hit in a game? Consider a game always to consist of 4 at-bats, and assume that at-bats are independent.

    b. What is the probability that he or she will get no hits in a game?

2. Conduct 60 trials in which you try to achieve the longest hitting streak possible for a hitter whose batting average you choose. Use a random number table to conduct the trials. Round the probability of getting a hit to two digits. If a value greater than or equal to your calculated probability comes up, the streak ends. Write the number of games that the streak lasts in the table below.

Trial	Games	Trial	Games	Trial	Games	Trial	Games	Trial	Games	Trial	Games
1		11		21		31		41		51	
2		12		22		32		42		52	
3		13		23		33		43		53	
4		14		24		34		44		54	
5		15		25		35		45		55	
6		16		26		36		46		56	
7		17		27		37		47		57	
8		18		28		38		48		58	
9		19		29		39		49		59	
10		20		30		40		50		60	

3. a. What was your batter's longest hitting streak?

    b. What was your batter's average hitting streak (number of games)?

         Real-Life Math: Probability

19. Two People in a Group With the Same Birthday

Context

coincidences

Math Topic

data collection and analysis

Overview

Unexpected coincidences often draw gasps. A famous problem in probability—calculation of the chance that at least two people in a small group share a birthday—illustrates that some types of coincidences are not at all unlikely.

Objectives

Students will be able to:

- determine through simulation the probability of at least two people in a group having a common birthday

Materials

- one copy of the Activity 19 handout for each pair of students

- for each class member, a set of two small paper bags, with one bag containing month names and the other containing dates obtained by cutting up copies of the Random Date Template from the Appendix

- calculator

Teaching Notes

Allow students to guess how often at least 2 people in a random group of students in a class your size will have the same birthday. Some people tend to guess that this is a rare occurrence because 367 people are required for a group guaranteed to have at least 2 members sharing a birthday.

This simulation will require the Random Date Template found in the Appendix. The instructions here assume that the simulation will be a whole-class project. The sets of month and date bags should be distributed to each member of the class.

For each trial, each student should shake up both bags, then draw out a single slip from each bag. If a student draws a date that does not exist for the drawn month, the student should draw another date without replacing the first slip. February 29 should be allowed, even though this birthday is less likely than any other. After the trial is recorded, all the slips must be returned to the proper bags, and the bags must be shaken up again. Just for fun, students may use their actual birthdays instead of drawn ones for one trial.

Students should report their drawn birthdays to a recorder at the board or at an overhead projector. If after all birthdays are reported, no two match, the trial should be recorded as "fail." If any two students

(continued)

19. Two People in a Group With the Same Birthday

reported matching birthdays, the trial should be recorded as "success."

Do as many trials as time allows. There are 30 spaces on the record sheet.

Answers

1. a–b. Answers will vary. If students are reluctant to make a prediction, throw out some proposals. For example, say, "How many think 1%?"

2. Results will vary.

3. a. Here are all the theoretical probabilities of at least one pair of common birthdays for group sizes from 2 to 35 students:

2, 0.3%	14, 22.3%
3, 0.8%	15, 25.2%
4, 1.6%	16, 28.3%
5, 2.7%	17, 31.4%
6, 4.0%	18, 34.6%
7, 5.6%	19, 37.8%
8, 7.4%	20, 41.1%
9, 9.4%	21, 44.3%
10, 11.7%	22, 47.5%
11, 14.1%	23, 50.6%
12, 16.7%	24, 53.7%
13, 19.4%	25, 56.8%

26, 59.7%	31, 72.9%
27, 62.6%	32, 75.2%
28, 65.3%	33, 77.4%
29, 68.0%	34, 79.4%
30, 70.5%	35, 81.3%

Note that for a group of 23, the probability is about 50%. For 35, it is more than 80%.

   b. Johnny Carson's experiment failed because he asked the audience if anyone had a particular birthday. In this experiment, students test for any two people having the same birthday on any date.

Extension Activity

Develop the mathematical model for the probabilities of common birthdays.

19. Two People in a Group With the Same Birthday

Some years ago on *The Tonight Show*, a guest tried to explain that there is a high probability that two people in a small group will have the same birthday. Then host, the late Johnny Carson, asked the audience of about 150 if anyone had his birthday. No one did. But the guest was not a mathematics student and couldn't find the flaw in Johnny's test. In this activity, you will perform an accurate test of the guest's statement.

1. a. How many students will participate in your group?

    b. Make a prediction—in what percentage of trials do you think there will be at least one pair of common birthdays?

2. You have been given two small bags to use in your trial. One bag contains the names of the months of the year. The other contains dates. To perform trials, you will draw random birthdays as follows for each trial.

    First, shake up the slips in your bags. Draw only one slip from each bag. One of the slips is your month, the other is your date. When it's your turn, report this birthday to the class. If you draw a date that doesn't exist for the month you draw—for example, September 31—draw another date without replacing the first slip. Although a birthday of February 29 is less likely than any other, it should be allowed.

    After all class members have reported birthdays, determine if any two of them match. If there are any matches, report the trial as "success." If there are no matches, report the trial as "failure." Check the appropriate box on the table on the next page.

(continued)

19. Two People in a Group With the Same Birthday

Trial	Success	Failure	Trial	Success	Failure	Trial	Success	Failure
1			11			21		
2			12			22		
3			13			23		
4			14			24		
5			15			25		
6			16			26		
7			17			27		
8			18			28		
9			19			29		
10			20			30		
						Total		

3. a. What percentage of all trials were reported as "success" and what percentage were reported as "failure"? How does this compare with your original prediction?

    b. What was wrong with Johnny Carson's experiment?

20. Medical Testing

Context

health

Math Topic

conditional probability

Overview

This activity introduces a sometimes subtle process in probability—what happens to probability once some new or additional fact about a situation is known. This is called *conditional probability* and can be quite confusing. Here students study a real-life problem concerning medical testing. No simulation is included here. (In Activity 26, another counterintuitive example of conditional probability concerning the 3-door problem is examined. Activity 26 includes a simulation.)

Objectives

Students will be able to:

- demonstrate that in medical testing, even a seemingly high accuracy rate can lead to a high rate of false positives

Materials

- one copy of the Activity 20 handout for each student

- calculator

Teaching Notes

This activity is designed to get students thinking about conditional probability— if *A* happens, what is the probability of *B* given *A*?

- You may want to try a few simple examples with the class.

- Tell students you have rolled a pair of dice which add up to at least 11. Ask them what the probability is that you have, in fact, rolled a 12. The answer is $^1/_3$. Once it is known the sum is at least 11, there are only three outcomes possible—5 and 6, 6 and 5, or 6 and 6, all equally likely. One of the three has a sum of 12.

What is the probability that a card drawn is a face card? The answer is $^{12}/_{52}$ or $^3/_{13}$. What is the probability that a card drawn is a jack if you already know it is a face card? The answer is 1/3, because a jack is one of three possible face cards. How about the probability that a card is a face card if you already know it is a jack? The answer is 1, because all jacks are face cards.

It is very easy to jump to conclusions when an impressive number, for example a high accuracy rate, is given. Medical testing is a prime example. Many conditions are relatively rare. In question 3, the rate of occurrence is only 0.2% (200 out of 100,000). If a test on an individual is 99% accurate, the number of possible false

(continued)

20. Medical Testing

positives on a condition with a 0.2% occurrence rate leads to a shocking rate of false positives. The important lesson here is to ask questions and understand probability when dealing with medical tests!

If the students should ask, of the 99,800 who do not have the condition, 98,802 will test negative. Of the 200 who do have the condition, 2 will test negative (false negatives). Only 2 of 98,804 or 0.002% will be false negative tests.

Answers

1. a. 100,000 – 200 or 99,800 do not have the condition.

    b. 200

    c. 0.99 × 200 or 198 of the 200 people who do have the condition will test positive, since the test is 99% accurate.

    d. 0.01 × 99,800 or 998 will falsely test positive since the test is 99% accurate.

    e. 198 + 998 = 1196 tests will be positive.

    f. 198 / 1196 = 0.166 or 16.6% of positive tests are true positives.

    g. 998 / 1196 = 0.834 or 83.4% of positive tests are false positives!

2. The result in this example that only 16.6% of positive tests are true positives is very surprising. An

impressive-sounding 99% accuracy rate can mean little if the test is for a relatively rare condition. Many medical tests are more useful for ruling out a condition, not proving that it exists. When you are dealing with medical tests, always ask your medical professional for rates of false positives! If a test comes back positive, always ask for confirmation from alternate tests or ask for another opinion.

Extension Activities

*   Ask students to investigate Bayes' Theorem. The calculations in these exercises all follow from this result first proved by Thomas Bayes in the eighteenth century.

*   Have students answer the questions in the activity again, this time assuming that the test is given twice to each person and that two positives are required to test positive.

20. Medical Testing

Should you be concerned about the accuracy of tests for medical conditions? For example, let's say a test for a medical condition is 99% accurate when it is given to one individual. Impressive, right? This means that on average, if 100 people are tested, 99 of the tests will give a correct result (negative or positive), and 1 will show an incorrect result. However, over a long time period, let's say that experience has shown that 200 out of 100,000 people in the entire population actually have this condition.

1. a. If 100,000 people are tested, how many of those 100,000 do not actually have the condition?

    b. How many of the 100,000 people actually do have the condition?

    c. Of the number of people who actually do have the condition, how many will test positive? Explain your answer.

    d. Of the number of people who do not have the condition, how many will falsely be shown by the test to be positive for the condition?

    e. What is the total number of positive tests out of the 100,000 tests administered?

    f. Given that a test is positive, what is the conditional probability that the person tested actually has the condition?

    g. Given that a test is positive, what is the conditional probability that the person tested does not have the condition (the test falsely shows a positive result)?

2. Comment on the results of these calculations of conditional probability. What questions should you ask a medical professional when you are tested for a medical condition?

21. Lotto Games: Winning the Big One?

Context

recreation

Math Topic

combinations

Overview

Participation in state-run lotteries has become a common recreational activity for millions of people. But how likely is a big lotto win? This activity gives students a chance to apply their knowledge of combinations and probability to develop a sense of reality about this question.

Objectives

Students will be able to:

- calculate lottery odds using combinations

- demonstrate a sense of the chance that an individual will win a lottery

Materials

- one copy of the Activity 21 handout for each student

- one copy of the Mock Lotto Tickets sheet (see Appendix) for each pair or small group of students

- calculator

Teaching Notes

This activity is based on the Florida Lotto game from the Florida state lottery. However, many other state lotteries operate similar games, and this activity could be modified to work with any of them.

Follow these steps:

- Prior to class, find the most recent winning number from Florida Lotto (or another lottery of your choice). This can easily be found on the Internet at www.flalottery.com.

- Students are first asked to analyze individually the total number of combinations of six distinct numbers that can be formed by choosing integers from 1 to 53. They will apply their knowledge from Activity 6 to perform this calculation. Students should calculate the total number of possible combinations and understand that the probability of a single ticket winning the big prize is 1/(total number of possible combinations).

- Next, have students organize into pairs or small groups. Distribute the Mock Lotto Tickets sheet to each group. This sheet contains 120 randomly generated combinations of six numbers picked from the integers 1 through 53.

(continued)

21. Lotto Games: Winning the Big One?

- Announce the winning number from the most recent Florida Lotto. The students should check the winning combination of numbers against all the combinations on the Mock Lotto Tickets. They should report how many mock combinations match 0, 1, 2, 3, 4, 5, or 6 of the numbers in the winning combination using the bar chart provided.

Students will find that they will rarely record a ticket with more than three matching numbers. The last question gives them the opportunity to comment on the reality of winning lottery games.

Answers

1. The total number of combinations is 53! / 47!(6!) = (53 × 52 × 51 × 50 × 49 × 48) / (6 × 5 × 4 × 3 × 2 × 1) = 22,957,480. The probability of winning on a single ticket is 1/22,957,480 or 0.0000000436. (Note that you would have to play almost 12 million individual picks in order to have the same probability of winning as calling a coin flip.)

2. Theory suggests that of the 120 mock tickets, 56 will match none of the numbers from the winning combination, 48 will match 1 of the numbers, 14 will match 2, and 2 will match 3. Note that is a total of 120. Tickets for which there will be more

than three matching numbers will be rare. Furthermore, actual results will almost never be an exact match to the theoretical values. There are far too few mock tickets for results to converge as suggested by theory, although rough correspondence should appear for zero, one, and two matches.

3. Because even three matched numbers is unusual, students may fairly respond that nearly everyone loses their money by playing the lottery.

Extension Activities

- This activity might be extended over several weeks and multiple drawings.

- Some students might be able to write computer programs to generate and test mock tickets.

63

21. Lotto Games: Winning the Big One?

Answer the following questions about the Florida lotto.

1. How many possible combinations of 6 distinct integers may be chosen from the integers 1 through 53?

    This is the total number of possible combinations of numbers that may be chosen in the Florida Lotto. Many other state lotteries have similar numbers of possibilities. What is the probability of a single ticket winning the Florida Lotto?

2. Write the most recent winning number combination from the Florida Lotto here:

    Work with a partner or a small group. Check each combination of 6 numbers from the Mock Lotto Tickets sheet and record the number of tickets having 1, 2, 3, 4, 5, or 6 numbers that match numbers in the winning combination. Make a bar chart to tabulate the results.

3. Your ticket must match three numbers to win a prize. What is the usual outcome of playing a state lotto game?

22. Waiting for the Academy Award®

Context

entertainment

Math Topic

waiting time

Overview

Few actors get a chance to act in big-name movies. But after an actor does make it big, what is the probability that the actor will win an Academy Award® (Oscar®) in any given year, or over an entire career? Some famous actors have waited decades and have appeared in dozens of movies before winning an Oscar®, while others have won for their very first movie. In truth, the probability of an actor winning one of the top awards for movie acting in any year is small.

Objectives

Students will be able to:

- estimate probabilities through rough assumptions

- simulate an individual actor's chance of winning an Oscar®

Materials

- one copy of the Activity 22 handout for each student

- one random number table for each student (see Appendix)

Teaching Notes

All assumptions about probability in this activity are extremely rough. The estimate of 200 available actors is based on the list of entertainment personalities and notable movies in the *World Almanac and Book of Facts 2006*. Students could investigate this for themselves.

There is no real reason to think that all movie actors will have an equal probability of winning each year. Many will be up in popularity, others down. Some will be more active, others less so. And the kinds of movies produced change every year, making some actors more desirable. However, the assumption of equal probabilities provides the easiest method for analyzing this simple model.

The simulation in the second part shows that for even a very talented actor to win an Oscar®, many things must fall into place. Good luck has much to do with it. Students are asked to design an experiment in which they pretend to be actors themselves. Give half the students Random Number Table 1, the other half Random Number Table 2 (these tables are in the Appendix). Ask the students to start at different positions in the tables.

A trial would go like this—the first probability (getting a good role) is $\frac{1}{2}$; say that a random number from 50 to 99 would mean success and 00 to 49 would mean failure. The second and third probabilities

(continued)

22. Waiting for the Academy Award®

(getting a nomination and getting the award) are each $1/5$, so for second and third random numbers, numbers from 80 to 99 would mean success and from 00 to 79 would mean failure. To win an award in any year, all three must be successes.

Students should keep careful count of the three random numbers for each year and should record the years (if any) in which they win. Just for fun, let students make up names of movies and short descriptions of the roles for which they win or are nominated.

Answers

1. $1/50$, assuming that the pool of actors is half male and half female

2. a. Student results will vary. There is a $1 - (0.98)^{50}$, or 64%, probability that an actor will win the award before the fifty-first year.

    b. See Teaching Notes.

Extension Activity

Students could investigate past Academy Award winners and the number of years into their careers that they received awards. From this research, they could study factors that improve an actor's probability of receiving an Oscar.

22. Waiting for the Academy Award®

The crowning achievement for an actor's career is winning an Academy Award (an Oscar). There are four acting awards each year—Best Actor in a Leading Role, Best Actress in a Leading Role, Best Actress in a Supporting Role, and Best Actor in a Supporting Role. But how many actors ever achieve this honor?

Some actors had decades of great performances before winning. Morgan Freeman finally won Best Actor in a Supporting Role for *Million Dollar Baby* in 2004 after acting in movies for over three decades. Robin Williams won the same award in 1997 for his role in *Good Will Hunting* after appearing in over 30 movies. And the late Jessica Tandy acted in dozens of movies over her 60-year career before winning a Best Actress Oscar for *Driving Miss Daisy* in 1989.

On the other hand, some actors have won early in their careers. For example, Jennifer Connelly won Best Actress in a Supporting Role for her first big role in *A Beautiful Mind*. Cuba Gooding, Jr., won for one of his first major films, *Jerry McGuire,* in 1996. And perhaps most remarkable, Anna Paquin at the age of 11 won the 1993 Oscar for Best Actress in a Supporting Role for *The Piano.*

1. A rough estimate of the number of movie actors and actresses in any given year that have the necessary credentials, have enough popularity, and are available to act in a film that might result in an Oscar win is 200. If all 200 available actors (assume equal numbers of male and female actors) have an equal probability of winning an award, what is the probability of any one of these actors winning an award in a given year?

2. Imagine that you are an actor just beginning your career in big movies. You are one of the 200 "available" movie actors. Assume that your chances of being in a movie with a role good enough to get an acting award are 1 in 2. Assume that the probability that you will be nominated if you get a role is $1/5$ and the probability you will win if nominated is $1/5$.

    a. Use a random number table to simulate your career of 50 years. Design and conduct an experiment in which you select three different random numbers for each year. If all three meet your success criteria, you win an Academy Award for that year. List the years (if any) that you win an award.

    b. In this activity, are the assumptions about probability good ones? Explain why or why not.

         Real-Life Math: Probability

23. Test for ESP

Context

paranormal phenomena

Math Topic

hypothesis testing

Overview

Belief in extrasensory perception (ESP) and other paranormal phenomena is rampant. Before we dismiss such beliefs, we should offer some methods for studying them. Any experiment designed to test for ESP will involve analyzing probabilities and gathering data.

Objectives

Students will be able to:

- conduct an experiment that tests for the presence of ESP

- develop a statistical basis for accepting or rejecting the presence of ESP

Materials

- one copy of the Activity 23 handout for each student

- one standard deck of 52 playing cards (no jokers) for each pair of students

Teaching Notes

Hypothesis testing is the crux of statistical analysis. Do data reveal a pattern that could be explained by a cause other than chance? The answer to this question is not always clear. In this activity, students will be guided through an experiment testing for ESP. They will use what they know about probability and statistics to develop reasons for accepting or rejecting the hypothesis that ESP exists.

The experiment is very simple: The subject states the suit of a randomly drawn playing card without looking. The result, correct or incorrect, is recorded. The experiment is repeated 10 times.

The interesting part of this activity lies in deciding how many right out of 10 will constitute a positive test for ESP.

The acceptance level will be the number of correct answers above which the results will not be attributed to random guessing alone. However, students should recognize that every outcome, from 0–10 successes, has a finite probability of occurring by chance.

The model is binomial, with each independent event having a probability of 1/4 (the chance of picking out one of four suits correctly entirely by chance). This model was discussed in Activity 9. The mathematical method for calculating these probabilities is as follows:

$$P_{n,m} = \binom{n}{m} p^m q^{(n-m)}$$

(continued)

23. Test for ESP

m	0	1	2	3	4	5	6	7	8	9	10
$P_{10,\,m}$	0.056	0.188	0.282	0.250	0.146	0.058	0.016	0.003	< 0.001	< 0.001	< 0.001

In this case, $p = \frac{1}{4}$, $q = \frac{3}{4}$, $n = 10$, and m runs from 0 to 10. The results of these calculations are displayed in the table above.

After students have all results from the trials, help them decide on the acceptance level. Make it clear that the acceptance level is up to the researcher. This is an issue in any kind of study. For an acceptance level of 7, the total probability that the subject will get 7 or more correct is only 0.003. An acceptance level of 6 means that there is a 1.9% chance that the subject will guess 6 or more correctly only by chance.

Answers

1. Results will vary. The table should be complete.

2. The acceptance level is up to the student. But make sure students are guided by the discussion above. For example, students should learn that selecting an acceptance level of 2 would yield over 50% incorrect acceptance that ESP is present.

Extension Activities

- Have students design other experiments of ESP on their own.

- Hand out a copy of experimental results from a scientific journal, and discuss how scientists use hypothesis testing to "prove" their hypotheses.

23. Test for ESP

Extrasensory perception (ESP) is an ability to detect, predict, or perceive events without the use of the usual senses (sight, hearing, touch) and without any prior knowledge. Other names for ESP include clairvoyance, telepathy, and precognition. Some people claim to have this power. Persons claiming they have ESP are claiming to have a paranormal ability. How can you test for ESP?

1. Get together with a partner and set up the following experiment. Imagine that one of you claims to have ESP. The other will administer a simple test to see if the claim is true. The tester shuffles a deck of regular playing cards, then selects a card from the middle of the deck without showing the card to the subject. The subject, without seeing the card, declares its suit—clubs, diamonds, hearts, or spades. Record the result by placing an X under "Success" or "Failure" in the table below. Repeat the trial 10 times.

Subject name _____   Tester name _____

Trial	Suit declared by subject	Suit actually drawn	Success	Failure
1				
2				
3				
4				
5				
6				
7				
8				
9				
10				
Total number of successes in 10 trials				

2. a. At what number of total successes out of 10 trials would you say your subject has ESP?

   b. Does your subject have ESP?

24. Hold Time

Context

telecommunication

Math Topic

queues

Overview

So far all the examples in this book have involved discrete (i.e., distinct) outcomes. For example, the toss of a coin has two distinct outcomes.

In this activity, the notion of a continuous distribution of outcomes is explored. Continuous random variables are particularly useful for problems in which events are happening over randomly varying intervals of time. Many such applied examples may be modeled by what is called a "Poisson process." The mathematical details of this process are beyond the scope of this book. Many textbooks are available that cover the topic.[1]

Certain real-life problems have properties of a Poisson process. Cars passing a point on a road, clicks of a Geiger counter measuring radioactive decay, and the arrival of customers at a service desk are all examples.

When customers arrive at a point of service, the rate they arrive and how fast they are served determines how long they will have to wait. In mathematics, the study of time waiting in line is called queuing

theory. This activity considers a queuing problem that is becoming all too common in daily life—holding on the telephone for a customer service representative. What's the probability that your service question will have to wait longer to be answered?

Objectives

Students will be able to:

- simulate a simple queuing model

Materials

- one copy of the Activity 24 handout (instructions and chart) for each student

- one random number table for each pair of students (see Appendix)

- calculator

Teaching Notes

The mathematics involved in analyzing a queuing model from a theoretical perspective will not be developed here. Even the professionals in this business often prefer a simulation over trying to crank out a lot of equations.

Two students should conduct the simulation. There is a wide range of possible ways to define the simulation. One possible set of procedures is described on the next page. Assume there are enough customers so that on average, 20 calls come in per hour.

(continued)

[1]See, for example, Sheldon M. Ross, *Introduction to Probability Models,* 8th ed. (San Diego, CA: Academic Press, 2002).

24. Hold Time

1. Each trial represents 1 minute. A trial begins with the first student checking the next random number in the table. If the number is between 00 and 32, there is a call (one call every 3 minutes is the average). For any other number, mark open lines *O* in the chart during that minute.

2. If there is a call, Student 2 will "answer" if a representative is available and use the next random number to determine the time of the conversation. Divide that next random number by 10, round up to the nearest whole number, and add 3 to get this time. Fill this number of minutes with *X*'s in the row for that available representative. An *X* in subsequent minutes then means that a representative is not available.

3. If a call comes in during a minute when both representatives have an *X*, then that customer goes on hold. Students should mark, count, and write down the number of hold minutes until a representative becomes available, then mark *X*'s for the length of the conversation with the customer coming off hold as described above.

4. Have students keep a tally of the number of calls that are lost because both representatives and the hold line are all busy.

You can instruct students to vary the parameters—average rate of calls per hour, range of possible call times—and try the simulation as many times as you like.

Here is one additional suggestion. Initially, the charts will be confusing. Do some dry runs with the entire class until students catch on to how to use them. This will reduce waste of charts.

Answers

The results will vary, depending on the parameters that are chosen and which rows of the random number tables are used. The sample answers here were obtained with 20 calls per hour and call times varying randomly between 4 and 13 minutes.

On the next page is a sample of how the record sheet might be used.

(continued)

24. Hold Time

1st Rep	X	X	X	X	X	X	X	X	X	X	X	X	X	X	X	X	X	X	X	X	X	X	X	X	X	X		
2nd Rep	0	0	0	X	X	X	X	X	X	X	X	X	X	X	X	X	X	X	X	X	X	X	X	X	X	X	4	13
On Hold	0	0	0	0	0	0	0	X	0	X	X	X	X	X	X	X	X	X	X	0	X	0	X	X				
1st Rep	X	X	X	X	X	X	X	X	X	X	X	X	X	X	X	X	X	X	X	X	X	X	X	X	X			
2nd Rep	X	X	X	X	X	X	X	X	X	X	X	X	X	X	X	X	X	X	0	0	0	0	0	0			4	15
On Hold	X	X	0	X	X	X	X	X	0	0	0	X	X	0	0	X	X	X	0	0	0	0	0	0				

1. 3 minutes for a full chart

2. 2 calls for a full chart

3. Opinions may vary here. In the sample chart, both representatives were busy nearly all the time, so yes, they would seem to be busy enough.

4. Again, a valid opinion would seem to be yes, given the sample parameters. The representatives were busy nearly all of the time, customers were on hold for what seems like a reasonable average of 3 minutes, and few calls were being lost.

24. Hold Time

You call your cell phone provider when you have a question about your bill. Your heart sinks when music comes on and a voice tells you that "all representatives are assisting other customers." It might be a long wait for service.

A company probably would go broke if it hired so many telephone representatives that no customer would ever have to be on hold while a lot of employees were sitting around with nothing to do. So, what is a reasonable hold time? How many representatives should a company hire so that the employees are busy most of the time?

In this activity, you will use a random number table to play a game that simulates calling customer service. You'll then figure out the average wait on hold. This is much easier than doing math equations to get the answer.

Here are the rules for your cell phone company's customer service:

1. There are two customer service representatives and one line on which a customer can be on hold.

2. Calls come in at an average rate of 20 per hour, but the exact times are random.

3. If someone calls when both representatives are busy and the hold line is occupied, that call is lost.

4. When a customer reaches a representative, they will stay connected for a random amount of time between 4 and 13 minutes.

To play, find a partner. One will use the random number table to decide when a customer is trying to call in. The other will be the "representative" and record the data on the supplied chart. Follow instructions given by your teacher about how to fill in the chart.

After your chart is complete, answer the following questions.

1. What is the average hold time after your chart is complete?

2. How many calls were lost?

3. Were your representatives busy enough? Explain.

4. In your opinion, does this company have the right number of customer service representatives? Explain your answer.

(continued)

24. Hold Time

Record Sheet for Customer Service Calls

Write "0" for available and "X" for unavailable/busy. Each block represents 1 minute; each row is 25 minutes.

	Totals Holds/Minutes
1st Rep	
2nd Rep	
On Hold	
1st Rep	
2nd Rep	
On Hold	
1st Rep	
2nd Rep	
On Hold	
1st Rep	
2nd Rep	
On Hold	
1st Rep	
2nd Rep	
On Hold	
1st Rep	
2nd Rep	
On Hold	
1st Rep	
2nd Rep	
On Hold	
1st Rep	
2nd Rep	
On Hold	
1st Rep	
2nd Rep	
On Hold	
1st Rep	
2nd Rep	
On Hold	
TOTAL minutes on hold, AVERAGE minutes on hold	

Real-Life Math: Probability

25. Parking at the Mall

Context

shopping

Math Topic

optimization

Overview

In a mall parking lot, customers arrive and leave at random intervals. People like to park near the door, so those close spaces have the lowest probability of being available. Students will devise the best strategies so that they waste the least amount of time in the parking lot.

Objectives

Students will be able to:

- develop a parking strategy through simulations

Materials

- one copy of the Activity 25 handout for each student
- one random number table for each student (see Appendix)

Teaching Notes

Although there are theoretical constructs for handling these kinds of "optimal stopping problems," simulations will be easier for students to understand. Parts 1a and 1b suggest strategies that students should test. Optional part 1c suggests that students should make up their own strategies.

The "cost" in this problem is the total amount of time spent searching for a space and then walking to and from the space. To evaluate the strategies, students will use random numbers to model the availability of parking spaces and will then calculate a cost function to see how well the strategy worked. For this exercise, the cost function is 5 (number of sections searched) + 2 (number of sections away from entrance). The object will be to find a strategy that minimizes the cost.

Students will use a random number table to check if a section is available. They should consider that parking is available if the sequential random number chosen is greater than or equal to the percentage representing the probability for that section. It will be helpful if students write down this list because the probabilities change from section to section.

Answers

1. The results will vary considerably. Here are some results from test trials:
   a. average cost function = 40
   b. average cost function = 36
2. Partners should compare results and report the best strategy.

Extension Activity

Go to a mall parking lot and make observations. Are the probabilities and cost function for this activity realistic? Refine these values and run new simulations.

25. Parking at the Mall

It happens every time you and your friends drive to the mall—you ignore the outer parking rows and drive through the first couple of rows nearest the door looking for a parking space. And there's always an argument.

"Why don't you take that one?" you say.

Your friend says, "No! We'll spend all our time walking from the car and back."

You say, "No, we'll spend all our time driving around the inner rows looking for a space we'll never get."

Try doing some simulations to develop a compromise strategy. Use a random number table to simulate the availability of a space in a section of the lot. Work with a partner.

1.   There are 25 parking sections at the mall. Section 1 is closest to the doors; Section 25 is farthest away. The probability of finding a parking place is 100% in Section 25, 96% in Section 24, and 92% in Section 23. The probability of finding a parking place in a section decreases by 4% per section as the sections get closer in, so that the probability is down to 4% by Section 1. You find that it takes 5 times longer to search a section by car than it does to walk a distance spanned by one section. So, in the trials below, the result will be a time cost equal to the time needed to search plus the time needed to walk to and from the mall entrance.

   a.   One partner should try this strategy: Search Section 1 first. If no space is found (a random number less than 96 appears), go on to Section 2. If no space is found (a random number less than 92 appears), go on to Section 3. Repeat until a space is found, and then record the time cost in the table below. Complete at least five trials.

   b.   The other partner should try an alternate strategy. For example, start with Section 12 (48% chance of finding a space). If no space is found, search Section 13. Repeat until a space is found, and then record the time cost below. Complete at least five trials. Use a separate sheet of paper to describe your strategy and present your results.

Strategy:	
Trial	Time cost = 5 (number of sections searched) + 2 (number of sections away from entrance)

26. The 3-Door Problem

Context

decision making

Math Topic

optimization

Overview

This problem, with its counterintuitive answer, should be a lively culmination of these activities. A complete statement of the problem is given on the student sheet.

Although similar problems have been discussed in textbooks for years,[1] it was *Parade Magazine* columnist Marilyn vos Savant who made this problem famous in 1991. For a thorough discussion of the problem and a summary of the amazing array of wrong answers submitted to vos Savant, even from the mathematics community, see the magazine *Skeptical Inquirer.*[2] Information about this problem can be located online at: http://mathworld.wolfram.com/MontyHall Problem.html.

Objectives

Students will be able to:

- apply careful analysis of conditional probability in order to make the best decision

- perform a simulation to test a hypothesis

Materials

- one copy of the Activity 26 handout for each student

Teaching Notes

Have students try to produce an answer for the problem before appealing to any theories of probability. The usual response is to say that it does not matter if you switch or not. Then, let students attempt to make their arguments based on probability theory, but do not correct them or give them any answers at this point.

Next, let them run the trials. The hint gives most of what they need to design the experiment. The students should be in pairs for the trials—one student is the host, the other the contestant. To use a random number table to generate a door for the prize, the host should move one digit at a time across the table. There are several

(continued)

---

[1] See B.W. Lindgren, G.W. McElrath, and D.A. Berry, *Introduction to Probability and Statistics,* 4th Edition. (New York: Macmillan, 1978), pp. 59–60.
[2] Gary P. Posner, "Nation's Mathematicians Guilty of 'Innumeracy,'" *Skeptical Inquirer,* Summer 1991, pp. 342–45; and Kendrick Frazier, "'Three Door' Problem Provokes Letters, Controversy," *Skeptical Inquirer,* Winter 1992, pp. 192–99

26. The 3-Door Problem

possible ways to use the digits. Perhaps let 1–3 represent Door 1, let 4–6 represent Door 2, and let 7–9 represent Door 3, but skip any 0's.

After the prize door is selected at random by the host, the host must not let the contestant know which door conceals the prize. The contestant then chooses an initial door and tells the host. At this point, the trial is defined because there are two possibilities—either the contestant initially chose the correct door, in which case she would win by staying with her choice, or the contestant initially chose the wrong door, in which case she would win by switching. Check off the appropriate box— "switch wins" or "stay wins." For the fun of it, students may go through having the host "show" the contestant the other nonwinning door and offering the switch.

Answers

The contestant should always switch. The theoretical answer is that the contestant will win ⅔ of the time by switching. The easiest way to see this is to consider the final result as only two separate outcomes—the contestant's original door and the set of 2 doors including the one known to conceal a goat and the one the contestant can switch to. The probability is ⅓ that the contestant originally chose the correct door at random. The conditional probability that the prize is behind the door that the contestant can switch to is ⅔ because the contestant has

already been shown by the host which of the 2 remaining doors not to choose.

There is an important point to remember about the trials—it is perfectly possible for 30 trials to yield values much different from the theoretical probability, even to the point where the "wrong" answer may look like the correct one. It would be a good idea to combine results from the whole class to get a better picture.

Extension Activity

Students may write a computer program to conduct thousands of trials of this model.

26. The 3-Door Problem

You are a contestant on a game show. The host lets you select 1 of 3 doors. Behind 1 door a great prize is hidden: an expensive sports car. You win the car if you choose the correct door. Behind the other 2 doors are junk prizes, "goats." First, you choose 1 door at random. But before showing if you have won, the host, who knows which door conceals the sports car, randomly opens 1 of the 2 doors you did not select, revealing a goat. The host then gives you a choice—stay with your original selection or switch to the other still-closed door.

1. To maximize your probability of winning the sports car, should you switch or not switch your choice of door?

2. Use what you know about probability to support your choice of door given the situation in question 1. What is the probability of winning if you switch doors? If you do not switch doors?

3. With a partner, design an experiment to test your answer to the 3-door problem. Conduct trials using a random number table and report your results in the table on the next page.

(continued)

26. The 3-Door Problem

Trial	Prize door	No. of switch wins	No. of stay wins	Trial	Prize door	No. of switch wins	No. of stay wins	Trial	Prize door	No. of switch wins	No. of stay wins
1				11				21			
2				12				22			
3				13				23			
4				14				24			
5				15				25			
6				16				26			
7				17				27			
8				18				28			
9				19				29			
10				20				30			

Hints: One of the partners is "host" and should use the random number table to decide which door the prize is behind. This partner must not tell the other partner, the "contestant," where the prize is located. The contestant then makes an initial choice of door. It is now possible to check off whether the contestant would have won by switching.

4. In what proportion of the trials did the contestant win by switching?

Random Number Table 1

35	84	81	49	54	56	85	83	31	54	58	16	72	15	80	31	22	60	53	76	62	10	78	50	80
62	91	36	17	37	20	12	67	30	86	65	35	34	36	75	41	51	56	56	45	13	67	76	76	97
29	34	72	29	76	63	39	31	55	34	75	34	31	51	95	91	88	10	08	61	49	17	37	16	82
45	45	14	25	28	53	57	90	67	07	13	86	83	28	41	80	63	32	89	72	91	07	45	42	19
25	92	30	85	73	30	04	86	36	33	04	77	87	08	56	95	51	59	87	19	51	70	91	38	45
18	30	35	88	44	20	26	70	55	19	07	34	17	26	22	86	47	78	41	37	82	86	73	03	24
20	64	71	05	02	72	02	41	42	73	18	17	06	66	75	07	21	94	59	95	24	26	77	18	39
87	23	49	64	09	17	20	28	68	21	31	44	22	21	61	54	10	84	23	63	43	26	92	12	47
42	56	54	72	56	10	76	84	29	84	56	03	05	68	18	65	66	22	00	65	78	55	72	47	93
78	43	85	76	08	47	28	47	46	19	33	22	78	58	44	65	61	43	77	34	36	12	53	69	75
84	85	22	13	25	94	47	75	27	21	48	75	73	94	87	27	14	65	36	27	93	22	17	39	57
09	68	85	74	02	13	17	06	42	73	93	85	00	01	82	82	30	72	79	96	06	60	03	35	69
05	86	59	95	55	09	50	90	24	21	11	03	60	94	22	80	73	94	90	51	74	29	04	66	84
10	15	38	39	23	69	24	20	68	02	80	32	41	23	18	21	08	83	32	95	77	44	92	65	19
46	05	05	96	22	51	70	04	33	65	76	66	50	15	76	10	56	82	49	25	74	29	80	40	82
82	16	83	09	25	41	12	22	32	65	25	11	81	03	97	32	80	50	49	44	44	04	20	62	33
05	68	98	67	61	08	75	72	68	51	20	11	38	00	43	22	77	70	20	72	67	35	17	30	67
15	90	19	40	91	18	67	37	55	46	36	35	32	48	13	18	83	47	89	53	46	72	95	64	63
38	64	15	17	37	87	78	15	66	33	79	80	97	87	70	58	48	65	17	00	59	51	95	86	35
91	73	13	40	95	45	08	37	98	20	84	08	72	25	85	15	44	94	03	63	75	92	98	64	18
35	35	52	14	21	56	17	21	29	84	65	11	08	74	36	07	65	75	10	12	70	13	28	12	44
61	22	25	21	43	15	77	69	52	15	40	32	59	13	48	11	06	85	74	41	70	40	58	67	02
58	27	52	67	21	33	35	74	42	51	89	22	41	94	27	73	63	06	73	56	97	72	69	02	42
08	23	95	25	06	80	64	31	12	95	78	77	43	51	16	55	29	29	84	88	25	87	02	85	03
64	16	69	68	87	82	16	79	99	88	68	96	93	84	49	47	97	74	54	84	99	40	02	18	92
46	67	87	95	80	92	09	63	15	90	05	39	86	42	34	96	05	71	39	35	91	52	23	02	96
70	23	95	73	26	05	13	27	28	80	53	26	18	31	58	02	58	66	77	93	25	46	67	69	73
63	34	40	53	10	50	42	28	05	32	47	75	14	21	29	97	32	83	65	20	85	80	17	82	00
66	64	23	53	78	02	66	16	91	12	53	77	81	62	31	40	67	00	60	72	52	10	83	53	34
86	28	15	03	10	29	73	59	17	32	32	32	15	11	58	43	29	21	81	55	79	79	77	69	04
23	93	87	65	32	78	86	52	81	78	68	93	80	64	41	21	94	30	01	76	58	52	87	54	02
38	99	15	95	74	57	83	63	91	10	00	62	73	31	33	37	97	40	06	16	25	04	22	07	85
82	95	67	51	57	46	07	07	18	80	85	05	48	70	42	90	61	95	77	34	55	98	15	60	18
88	08	67	55	15	33	56	18	14	44	43	71	45	04	16	37	29	20	25	57	02	64	76	20	25
56	03	61	58	76	55	57	92	24	87	05	60	88	01	23	94	05	10	76	67	57	85	80	76	04
96	51	12	09	26	40	04	93	01	30	72	33	62	13	64	39	09	26	92	46	71	42	18	10	35
11	60	81	89	07	74	73	56	39	27	19	63	84	32	80	15	26	57	13	98	07	37	96	68	90
89	75	18	36	19	39	69	03	52	69	31	07	05	77	44	28	87	63	59	80	46	11	45	33	74
16	63	29	36	97	57	46	90	47	22	35	43	50	45	83	87	33	13	08	22	71	27	49	63	22
46	79	42	25	31	81	36	75	58	33	46	14	95	12	46	72	79	23	21	71	52	43	47	23	38

Random Number Table 2

63	70	71	94	53	65	79	75	59	68	58	30	51	38	90	09	84	56	13	87	40	04	65	98	92
26	48	58	82	13	27	28	22	28	32	50	13	46	68	08	12	75	51	80	27	14	16	30	95	48
39	58	95	15	96	74	63	30	99	25	79	13	57	23	98	67	70	21	02	38	00	80	51	96	30
35	33	55	69	09	19	81	93	62	41	90	52	95	49	65	92	74	78	18	42	17	59	01	13	08
47	74	37	86	65	71	58	69	62	66	25	73	45	65	11	80	76	63	14	12	45	50	66	28	54
36	68	25	50	56	01	96	56	71	07	23	34	93	02	78	33	35	96	41	70	55	39	87	27	81
23	84	96	30	80	61	92	72	28	17	81	33	36	49	30	07	09	45	12	97	50	38	91	21	63
37	92	39	12	25	11	03	13	57	14	05	59	22	24	57	23	92	96	79	68	25	75	50	13	14
83	18	24	48	17	99	09	37	55	78	87	85	30	54	09	42	67	77	02	61	60	36	96	39	68
16	43	85	98	04	13	27	59	45	90	78	33	82	37	13	89	35	43	30	07	13	11	87	78	86
07	73	56	94	15	06	56	70	03	69	04	32	80	67	66	06	94	35	42	06	86	32	90	53	04
29	90	16	19	88	26	90	48	56	11	21	88	73	57	17	30	84	70	80	32	97	01	16	21	42
04	27	95	98	91	01	08	81	53	47	39	41	36	18	12	92	14	57	10	32	87	67	04	24	80
73	94	66	13	74	69	67	49	17	63	80	22	90	04	16	58	03	50	59	74	42	57	80	68	12
64	95	51	55	23	31	88	15	28	95	79	43	74	59	78	82	01	12	24	40	59	27	28	62	49
17	54	31	63	12	43	40	81	47	40	64	61	28	31	33	86	33	55	89	94	53	85	95	82	49
41	97	55	64	60	24	18	15	71	65	80	96	69	19	92	31	39	74	83	69	93	47	11	26	92
94	63	97	69	85	68	88	41	65	72	06	96	14	68	54	03	96	30	36	12	17	68	55	13	56
89	45	41	73	29	47	17	63	66	77	48	20	80	06	59	64	24	93	38	03	82	04	74	88	88
54	20	58	06	21	54	86	82	61	21	67	72	35	61	87	84	59	76	68	18	61	43	10	44	84
46	36	20	74	52	03	91	14	26	12	56	34	53	87	86	54	24	15	17	89	23	75	73	18	08
82	46	72	34	38	65	31	82	73	78	86	15	67	04	76	98	92	09	59	36	81	42	70	34	97
79	89	68	38	79	70	27	60	57	11	97	53	49	59	67	56	95	93	54	44	64	07	05	55	53
51	14	86	89	87	43	36	93	68	58	20	34	96	66	32	79	28	39	12	53	10	63	66	33	15
67	31	21	81	10	80	55	15	30	62	48	62	29	41	26	06	44	41	36	07	49	81	37	30	46
81	30	46	25	15	05	60	97	83	20	71	48	49	90	37	57	70	67	09	79	51	24	97	46	41
78	10	64	12	59	35	32	11	98	70	52	81	66	23	38	20	38	76	15	64	47	94	98	44	59
23	12	26	37	14	01	17	40	10	85	26	54	64	81	78	27	19	41	68	50	47	42	28	26	29
93	91	66	54	82	47	13	08	12	52	71	21	91	37	80	51	91	10	31	98	92	66	49	54	40
00	80	64	00	84	54	38	69	42	50	72	95	74	47	96	70	75	96	35	41	58	95	99	35	93
64	29	22	26	75	13	89	84	69	60	35	64	53	08	87	67	81	80	69	31	96	64	06	89	67
50	09	31	07	46	89	83	16	51	62	68	83	82	15	06	74	75	86	95	81	94	19	56	19	62
68	22	59	00	37	24	55	38	81	59	54	61	17	71	40	45	22	96	54	45	87	90	17	60	51
84	14	52	51	71	66	21	47	87	35	70	85	82	70	37	87	64	38	08	58	78	17	05	20	67
76	56	25	42	91	01	80	01	58	19	63	83	10	40	23	92	91	07	25	39	76	57	18	33	53
73	20	10	00	32	62	97	61	24	08	30	34	15	77	01	42	11	72	86	16	77	13	20	05	00
09	87	19	93	90	51	86	77	16	12	35	32	02	75	39	37	33	48	58	01	70	95	88	11	24
70	53	31	97	24	66	40	68	76	28	80	58	14	43	19	20	87	60	87	51	03	84	46	24	87
91	97	50	82	87	81	83	56	18	67	16	49	98	79	18	11	15	99	71	95	86	65	95	01	43
80	42	65	35	88	96	09	14	75	97	81	31	11	10	50	41	93	28	03	97	91	51	84	24	53

Mock Lotto Tickets

01	04	11	17	19	46	06	08	18	19	26	39	14	38	39	44	47	50
04	13	30	39	41	45	10	12	35	38	41	48	03	13	14	27	33	39
08	26	30	43	48	50	02	10	16	25	40	49	07	10	18	24	25	28
01	04	16	35	37	52	06	08	22	26	29	52	15	19	23	30	49	52
04	14	20	30	38	52	10	20	28	36	41	43	03	13	22	23	24	35
09	10	13	15	27	32	02	10	16	45	51	52	07	11	20	21	29	40
01	06	08	27	31	50	06	08	25	34	44	49	15	26	28	34	40	49
04	14	28	29	30	46	10	29	32	43	47	48	03	19	25	40	45	46
09	10	25	26	32	49	02	10	27	41	46	52	07	13	30	38	46	50
01	09	19	30	32	45	06	09	23	29	30	53	15	32	39	40	45	49
04	16	20	23	45	46	11	12	27	28	50	53	04	05	19	37	43	49
09	11	19	27	28	29	02	11	16	18	21	42	07	14	15	29	36	46
01	10	24	33	37	50	06	10	18	20	49	50	15	33	41	42	43	46
04	34	35	37	49	52	11	13	25	30	32	41	04	06	14	25	28	45
09	11	22	34	47	52	02	12	28	35	43	52	07	19	24	34	36	44
01	10	28	44	52	53	06	10	38	45	52	53	16	20	23	27	38	43
05	06	28	29	31	48	11	14	15	21	38	51	04	07	17	38	43	47
09	11	25	26	34	41	02	13	18	28	38	53	07	21	30	37	47	50
01	12	15	21	32	35	06	11	16	28	35	38	17	18	24	28	38	47
05	07	22	26	36	38	11	18	23	24	26	50	04	08	15	17	43	44
09	12	14	16	39	50	02	15	18	31	43	47	07	21	36	38	48	53
01	21	28	32	48	53	06	12	22	35	36	42	17	22	32	35	36	40
05	12	14	20	41	44	11	18	33	42	48	53	04	10	11	12	38	49
09	12	29	30	32	51	02	19	22	42	45	53	08	09	11	15	38	45
01	24	29	31	43	53	06	13	17	27	38	43	19	20	26	29	42	47
05	12	14	26	29	42	11	25	26	29	49	53	04	10	11	12	13	36
09	13	33	35	40	42	02	19	33	47	49	52	08	10	37	40	43	47
01	28	34	37	43	46	06	16	17	38	40	41	20	26	30	31	38	50
05	16	22	23	30	52	12	19	21	29	34	40	04	10	21	30	33	45
09	17	39	44	48	52	02	21	26	29	39	51	08	13	15	41	42	44
01	32	38	39	40	47	06	16	28	31	36	52	22	24	27	41	42	46
06	07	09	10	16	40	12	22	23	44	48	51	04	11	26	47	49	50
09	19	21	32	36	51	03	06	12	34	40	43	08	14	31	37	38	52
02	04	16	29	39	44	06	18	26	29	38	49	25	30	31	32	42	52
06	07	13	38	50	52	14	15	36	37	44	49	04	12	21	22	41	53
09	20	22	26	38	47	03	07	13	29	42	43	08	15	21	22	29	44
02	05	07	23	46	49	07	09	27	37	38	44	27	33	34	40	42	48
06	07	15	30	34	44	14	25	29	34	36	43	04	13	19	28	29	43
09	23	36	37	49	51	03	11	13	23	27	32	08	19	32	34	38	44
02	07	22	23	34	41	07	10	15	18	36	43	28	32	35	36	43	44

Random Date Template

January	February	March	April
May	June	July	August
September	October	November	December

1	2	3	4	5	6	7
8	9	10	11	12	13	14
15	16	17	18	19	20	21
22	23	24	25	26	27	28
29	30	31				

WALCH PUBLISHING

Share Your Bright Ideas

We want to hear from you!

Your name_____Date_____

School name_____

School address_____

City _____State _____Zip_____Phone number (_____)_____

Grade level(s) taught_____Subject area(s) taught_____

Where did you purchase this publication?_____

In what month do you purchase a majority of your supplements?_____

What moneys were used to purchase this product?

____School supplemental budget ____Federal/state funding ____Personal

Please "grade" this Walch publication in the following areas:

	A	B	C	D
Quality of service you received when purchasing	A	B	C	D
Ease of use	A	B	C	D
Quality of content	A	B	C	D
Page layout	A	B	C	D
Organization of material	A	B	C	D
Suitability for grade level	A	B	C	D
Instructional value	A	B	C	D

COMMENTS:_____

_____

What specific supplemental materials would help you meet your current—or future—instructional needs?

_____

Have you used other Walch publications? If so, which ones?_____

May we use your comments in upcoming communications? ____Yes ____No

Please **FAX** this completed form to **888-991-5755**, or mail it to

Customer Service, Walch Publishing, P. O. Box 658, Portland, ME 04104-0658

We will send you a **FREE GIFT** in appreciation of your feedback. **THANK YOU!**